EAA Lecture Notes

Editors

H. Bühlmann   A. Pelsser
W. Schachermayer H. Waters   D. Filipovic, Chair

EAA Lecture Notes is a series supported by the European Actuarial Academy (EAA GmbH), founded on the 29 August, 2005 in Cologne (Germany) by the Actuarial Associations of Austria, Germany, the Netherlands and Switzerland. EAA offers actuarial education including examination, permanent education for certified actuaries and consulting on actuarial education.

actuarial-academy.com

EAA Lecture Notes

Wüthrich, M.V.; Bühlmann, H.; Furrer, H. **Market-Consistent Actuarial Valuation** 2007

Eva Lütkebohmert

Concentration Risk
in Credit Portfolios

With 17 Figures and 19 Tables

Eva Lütkebohmert
Universität Bonn
Inst. für Gesellschafts- und
Wirtschaftswissenschaften
Adenauerallee 24-42
53113 Bonn
Germany
eva.luetkebohmert@uni-bonn.de

ISBN 978-3-540-70869-8   e-ISBN 978-3-540-70870-4

DOI 10.1007/978-3-540-70870-4

EAA Lecture Notes ISSN 1865-2174

Library of Congress Control Number: 2008936503

© 2009 Springer-Verlag Berlin Heidelberg

This work is subject to copyright. All rights are reserved, whether the whole or part of the material is concerned, specifically the rights of translation, reprinting, reuse of illustrations, recitation, broadcasting, reproduction on microfilm or in any other way, and storage in data banks. Duplication of this publication or parts thereof is permitted only under the provisions of the German Copyright Law of September 9, 1965, in its current version, and permission for use must always be obtained from Springer. Violations are liable to prosecution under the German Copyright Law.

The use of general descriptive names, registered names, trademarks, etc. in this publication does not imply, even in the absence of a specific statement, that such names are exempt from the relevant protective laws and regulations and therefore free for general use.

Cover design: WMXDesign GmbH, Heidelberg

Printed on acid-free paper

9 8 7 6 5 4 3 2 1

springer.com

Preface

The modeling and management of credit risk is the main topic within banks and other lending institutions. *Credit risk* refers to the risk of losses due to some credit event as, for example, the default of a counterparty. Thus, credit risk is associated with the possibility that an event may lead to some negative effects which would not generally be expected and which are unwanted. The main difficulties, when modeling credit risk, arise from the fact that default events are quite rare and that they occur unexpectedly. When, however, default events take place, they often lead to significant losses, the size of which is not known before default. Although default events occur very seldom, credit risk is, by definition, inherent in any payment obligation. Banks and other lending institutions usually suffer severe losses when in a short period of time the quality of the loan portfolio deteriorates significantly. Therefore, modern society relies on the smooth functioning of the banking and insurance systems and has a collective interest in the stability of such systems. There exist, however, several examples of large *derivative losses* like Orange County (1.7 billion US$), Metallgesellschaft (1.3 billion US$) or Barings (1 billion US$), which took place between 1993 and 1996. These examples have increased the demand for regulation aiming at financial stability and prove that risk management is indeed a very important issue.

A particular form of credit risk is referred to as *concentration risk*. Concentration risks in credit portfolios arise from an unequal distribution of loans to single borrowers (*name concentration*) or industrial or regional sectors (*sector* or *country concentration*). In addition, certain dependencies between different borrowers can increase the credit risk in a portfolio since the default of one borrower can cause the default of a dependent second borrower. This effect is called *default contagion* and is linked to both name and sector concentration.

Historical experience shows that concentration of risk in asset portfolios has been one of the major causes of bank distress. In the 1980s banks in Texas and Oklahoma suffered severe losses in both corporate and commercial real estate lending due to significant concentrations of lending in the energy industry. Moreover, the regional dependence on oil caused strong correlation between the health of the energy industry and local demand for commercial real estate. In the last years, the failures of large borrowers like Enron, Wordcom and Parmalat were the source of sizeable losses in a number of banks. Furthermore, the relevance of concentration risk is demonstrated by the recent developments in conjunction with the *subprime mortgage crisis,* which started in the United States in late 2006 and turned into a global financial crisis during 2007 and 2008. Subprime refers to loans granted to borrowers of low credit-worthiness. In the United States many borrowers speculated on rising housing prices and assumed mortgages with the intention to profitably refinance them later on. However, housing prices started to drop in 2006 and 2007 so that refinancing became increasingly difficult and, finally, lead to several defaults. Many mortgage lenders had used securitization methods as, for example, mortgage-backed securities (MBSs) or collateralized debt obligations (CDOs). Thereby the credit risk associated with subprime lending had been distributed broadly to corporate, individual and institutional investors holding these MBSs and CDOs. High mortgage payment defaults then lead to significant losses for financial institutions, accumulating to US$170 billion in March 2008. The crisis had spread worldwide. In Germany, for example, the IKB Deutsche Industriebank suffered dramatic losses from the subprime market downturn, leading to a bail out in August 2007. This also points out the importance of an effective measurement and management of concentration risk. Within the subprime or mortgage sector, loans are highly correlated and defaults spread quickly by contagion effects.

In addition to these examples, the amount of specific rules, which are imposed by the supervisory authorities to control concentration of risks, shows the importance of diversification of loan portfolios with respect to regions, countries and industries. These examples, rules and the ongoing research demonstrate that it is essential for the evaluation and management of a bank's credit risk to identify and measure concentration of risks within a portfolio.

It is the aim of these lecture notes to reflect the recent developments in research on concentration risk and default contagion both from an academic and from a supervisory perspective. After a short introduction to credit risk in general, we study in Part I several important model-based approaches which allow banks to compute a probability distribution of credit losses at portfolio level. Being the precursor of the asset-value models, the famous *Merton model* is discussed in some detail. Based on this approach we present some of the most prominent industry models to capture portfolio credit risk, namely *CreditMetrics* of J.P. Morgan [77] and *PortfolioManager* of KMV [84]. Moreover, as a representative of the class of *reduced-form* or *mixture models,* we discuss the *CreditRisk$^+$* model of Credit Suisse [30]. Besides these industry models, we also study the *Internal Ratings Based* (IRB) approach, on which

Basel II is based. These models then provide a basis for measuring concentration risks, which will be the main task of Part II. Here, we also give a short overview of simple ad-hoc measures for the quantification of concentration risk in credit portfolios. For a decent measurement of concentration risk, however, model-based approaches are preferable. Thus, most of Part II deals with different methods for the measurement of name and sector concentration in credit portfolios. Part III focusses on the topic of default contagion which is linked to both name and sector concentration risk. Here we can distinguish mainly three classes of approaches; the *copula models,* methods from *interacting particle systems* and *equilibrium models.* We give an overview over the different approaches for the treatment of default contagion and discuss the similarities between these methodologies. The book is intended to reflect the current state of research in the area of concentration risk and default contagion. While discussing potential drawbacks of some of the approaches, we also give an outlook in which direction research in this area can lead in future.

These lecture notes were developed when I gave a PhD course on financial economics at the University of Bonn. They are intended for mathematicians as well as for economists having a profound background in probability theory. I have included an introduction to credit risk modeling in Part I to keep this book as self-contained as possible. In Parts II and III, I focus on the main ideas behind the presented approaches. Therefore, I try to avoid some of the technicalities while maintaining mathematical precision. For detailed proofs of some of the statements I refer to the original papers presented.

I would like to thank the PhD students in my course, Marcelo Cadena, Daniel Engelage, Haishi Huang, Jördis Klar, Stefan Koch, Birgit Koos, Bernd Schlusche, Klaas Schulze and Manuel Wittke for their contributions to this work. Special thanks also go to Sebastian Ebert for proof-reading and for many helpful comments. I appreciate the stimulating discussions in the subgroup on concentration risks of the Basel Research Task Force in which I participated while working for the department of banking supervision of the Deutsche Bundesbank. The work in this group and, in particular, the joint project with Michael Gordy, also were the initiative for me to give a course on the topic of concentration risk and to write these lecture notes. In particular, I would like to thank Michael Gordy, Dirk Tasche and Klaus Düllmann for many stimulating discussions. I am particularly grateful to Rüdiger Frey and to two anonymous referees for detailed feedback and many helpful comments which significantly improved this book.

Bonn, May 2008   *Eva Lütkebohmert*

Contents

Part I Introduction to Credit Risk Modeling

1 Risk Measurement 3
   1.1 Variables of Risk 4
   1.2 The General Model Setting 5
   1.3 Exchangeable Models 7

2 Modeling Credit Risk 9
   2.1 The Regulatory Framework 10
   2.2 Expected and Unexpected Loss 12
   2.3 Value-at-Risk 13
   2.4 Expected Shortfall 15
   2.5 Economic Capital 17

3 The Merton Model 19
   3.1 The General Framework 20
   3.2 The Multi-Factor Merton Model 23
   3.3 Industry Models Based on the Merton Approach 29
      3.3.1 The KMV Model 29
      3.3.2 The CreditMetrics Model 30

4 The Asymptotic Single Risk Factor Model 31
   4.1 The ASRF Model 32
   4.2 The IRB Risk Weight Functions 35
   4.3 The Loss Distribution of an Infinitely Granular Portfolio 38

5 Mixture Models 43
   5.1 Bernoulli and Poisson Mixture Models 43
   5.2 The Influence of the Mixing Distribution on the Loss
      Distribution 48
   5.3 Relation Between Latent Variable Models and Mixture Models 50

6 The CreditRisk$^+$ Model 53
6.1 Basic Model Setting 54
6.2 The Poisson Approximation 56
6.3 Model with Random Default Probabilities 57

Part II Concentration Risk in Credit Portfolios

7 Introduction 63

8 Ad-Hoc Measures of Concentration 67
8.1 Concentration Indices 68
8.2 Conclusion 72

9 Name Concentration 75
9.1 A Granularity Adjustment for the ASRF Model 76
9.1.1 Example as Motivation for GA Methodology 77
9.1.2 The General Framework 78
9.1.3 The Granularity Adjustment in a Single Factor CreditRisk$^+$ Setting 81
9.1.4 Data on German Bank Portfolios 84
9.1.5 Numerical Results 86
9.1.6 Summary 88
9.2 The Semi-Asymptotic Approach 90
9.2.1 The General Framework 90
9.2.2 Numerical Results 93
9.3 Methods Based on the Saddle-Point Approximation 93
9.3.1 The General Framework 94
9.3.2 Application to Name Concentration Risk 96
9.4 Discussion and Comparison Study of the Granularity Adjustment Methods 99
9.4.1 Empirical Relevance of the Granularity Adjustment 100
9.4.2 Why a Granularity Adjustment Instead of the HHI? ... 100
9.4.3 Accuracy of the Granularity Adjustment and Robustness to Regulatory Parameters 102
9.4.4 Comparison of Granularity Adjustment with Other Model-Based Approaches 103
9.4.5 Agreement of Granularity Adjustment and Saddle-Point Approximation Method in the CreditRisk$^+$ Model 104

10 Sector Concentration 107
10.1 Analytical Approximation Models 108
10.1.1 Analytical Approximation for Value-at-Risk 109
10.1.2 Analytical Approximation for Expected Shortfall 117
10.1.3 Performance Testing 118

          10.1.4 Summary and Discussion 119
     10.2 Diversification Factor Models 120
          10.2.1 The Multi-Sector Framework 121
          10.2.2 The Capital Diversification Factor 123
          10.2.3 Marginal Capital Contributions 124
          10.2.4 Parameterization 126
          10.2.5 Application to a Bank Internal Multi-Factor Model 127
          10.2.6 Discussion 129

11 **Empirical Studies on Concentration Risk** 131
     11.1 Sector Concentration and Economic Capital 132
          11.1.1 The Model Framework 133
          11.1.2 Data Description and Portfolio Composition 133
          11.1.3 Impact of Sector Concentration on Economic Capital .. 135
          11.1.4 Robustness of EC Approximations 136
          11.1.5 Discussion 139
     11.2 The Influence of Systematic and Idiosyncratic Risk on Large
          Portfolio Losses 140
          11.2.1 Descriptive Analysis of SNC Data 140
          11.2.2 Simple Indices of Name and Sector Concentration 141
          11.2.3 Modeling Dependencies in Losses 142
          11.2.4 Monte Carlo Simulation of the Portfolio Loss
                 Distribution 143
          11.2.5 Empirical Results 145
          11.2.6 Summary and Discussion 147

Part III Default Contagion

12 **Introduction** 151

13 **Empirical Studies on Default Contagion** 155
     13.1 The Doubly Stochastic Property and its Testable Implications 156
     13.2 Data for Default Intensity Estimates 159
     13.3 Goodness-of-Fit Tests 159
     13.4 Discussion 162

14 **Models Based on Copulas** 165
     14.1 Equivalence of Latent Variable Models 166
     14.2 Sensitivity of Losses on the Dependence Structure 168
     14.3 Discussion 170

XII   Contents

15 A Voter Model for Credit Contagion 173
    15.1 The Model Framework 174
    15.2 Invariant and Ergodic Measures for the Voter Model 177
    15.3 The Non-Dense Business Partner Network 179
    15.4 The Dense Business Partner Network 180
    15.5 Aggregate Losses on Large Portfolios 182
    15.6 Discussion and Comparison with Alternative Approaches 186
        15.6.1 The Mean-Field Model with Interacting Default
               Intensities 187
        15.6.2 A Dynamic Contagion Model 189
    15.7 Contagion Through Macro- and Microstructural Channels 190
        15.7.1 A Model with Macro- and Micro-Structural Dependence 191
        15.7.2 The Rating Migrations Process 193
        15.7.3 Results and Discussion 194

16 Equilibrium Models 197
    16.1 A Mean-Field Model of Credit Ratings 198
    16.2 The Mean-Field Model with Local Interactions 202
    16.3 Large Portfolio Losses 205
    16.4 Discussion 208

A   Copulas 211

References 217

Index 223

Abbreviations

ASRF	asymptotic single risk factor
CDI	capital diversification index
CDO	collateralized debt obligation
CDS	credit default swap
CGF	cumulant generating function
CP2	second consultative paper
DD	distance-to-default
DF	diversification factor
EAD	exposure at default
EC	economic capital
EDF	expected default frequency (KMV model)
EL	expected loss
ES	expected shortfall
GA	granularity adjustment
HHI	Herfindahl-Hirschman Index
iid	independent identically distributed
IRB	internal ratings based
KMV model	credit risk model by Moody's KMV
LGD	loss given default
MA	maturity adjustment
MC	Monte Carlo
MCR	minimum capital requirements
MFA	multi factor adjustment
MGF	moment generating function
PD	probability of default
pdf	probability density function
PGF	probability generating function
RWA	risk weighted assets
UL	unexpected loss
VaR	value-at-risk

Notations

General Notations

$\mathbb{P}$	probability measure
$\mathbb{E}[\cdot]$	expectation with respect to probability measure $\mathbb{P}$
$\mathbb{V}[\cdot]$	variance with respect to probability measure $\mathbb{P}$
$G_X(s)$	probability generating function of a random variable X
$M_X(s)$	moment generating function of a random variable X
$K_X(s)$	cumulant generating function of a random variable X
Φ	cumulative standard normal distribution function
ϕ	density of standard normal distribution
$\Phi_d[\cdot;\Gamma]$	d-dimensional cumulative centered Gaussian distribution function with correlation matrix Γ
t_ν	univariate Student t-distribution function with ν degrees of freedom
$t_{\nu,R}^d$	d-dimensional multivariate t-distribution function with ν degrees of freedom and correlation matrix R
$\mathrm{Bern}(1;p)$	Bernoulli distribution with event probability p
$\alpha_q(X)$	q^{th} quantile of random variable X
$\mathrm{VaR}_q(X)$	value-at-risk at level q of random variable X
$\mathrm{ES}_q(X)$	expected shortfall at level q of random variable X
EC_q	economic capital at level q
N	number of obligors in a portfolio (unless otherwise stated)

Obligor Specific Quantities

EAD_n	exposure at default of obligor n
s_n	exposure of obligor n as share of total portfolio exposure
PD_n	default probability of obligor n
LGD_n	loss given default of obligor n
ELGD_n	expectation of loss given default variable of obligor n
VLGD_n	volatility of loss given default variable of obligor n

D_n default indicator variable of obligor n
U_n loss rate on obligor n, equal to $\text{LGD}_n \cdot D_n$
$\mathcal{R}_n$ EL capital requirement of obligor n
$\mathcal{K}_n$ UL capital requirement of obligor n

Single Factor Models

L resp. L_N	loss variable of portfolio with N loans
X	systematic risk factor
ε_n	idiosyncratic shock of obligor n
$\mu(X) = \mathbb{E}[L_N\|X]$	conditional mean of the portfolio loss
$\sigma^2(X) = \mathbb{V}[L_N\|X]$	conditional variance of the portfolio loss
$\mu_n(X) = \mathbb{E}[U_n\|X]$	conditional mean of the loss ratio of obligor n
$\sigma_n^2(X) = \mathbb{V}[U_n\|X]$	conditional variance of the loss ratio of obligor n

Multi Factor Merton Model

$V_t^{(n)}$	asset value of obligor n
r_n	asset value log returns of obligor n
C_n	threshold value of obligor n corresponding to $V_T^{(n)}$
c_n	threshold value of obligor n corresponding to r_n
L_n	loss variable of obligor n
L	portfolio loss variable
ε_n	idiosyncratic shock of obligor n
K	number of systematic risk factors
$X_1, \ldots, X_K$	independent systematic risk factors
Y_n	composite factor of obligor n
β_n	factor loading describing the correlation between r_n and Y_n
$\alpha_{n,k}$	weight describing the dependence of obligor n on factor X_k

ASRF Model

L_N	portfolio loss variable for a portfolio of N obligors
LGD_n	downturn LGD of obligor n
r_n	asset value log returns of obligor n
ϱ_n	asset correlation of obligor n

Mixture Model

$K < N$	number of factors
$X = (X_1, \ldots, X_K)$	factor vector
D_n	default indicator variable of obligor n in Bernoulli mixture model
$\tilde{D}_n$	number of defaults of obligor n up to time T in Poisson mixture model

M	number of defaulted companies in Bernoulli mixture model
$\tilde{M}$	approximate number of defaulted companies in Poisson mixture model
$p_n(X)$	function depending on the factor vector X, describing the Bernoulli parameter (in exchangeable case $p_n(X) = p(X)$)
$\lambda_n(X)$	function depending on the factor vector X, describing the intensity parameter in the Poisson mixture model (in exchangeable case $\lambda_n(X) = \lambda(X)$)
G_p	probability distribution function of random variable $p(X)$
G_λ	probability distribution function of random variable $\lambda(X)$

CreditRisk$^+$ Model

E_n	loss exposure of obligor n, equal to $\text{EAD}_n \cdot \text{ELGD}_n$
E	base unit of loss
D_n	default indicator of obligor n
PD_n	default probability of obligor n (possibly random, otherwise the average PD)
$\text{PD}_n(X)$	random default probability of obligor n, depending on X
L_n	loss variable of obligor n
L	portfolio loss variable
ν_n	normalized exposure of obligor n, equal to E_n/E
λ_n	normalized loss of obligor n, equal to $D_n \cdot \nu_n$
λ	normalized portfolio loss
K	number of systematic risk factors
$X_1, \ldots, X_K$	independent Gamma distributed systematic risk factors
$X_0 = 1$	idiosyncratic risk
$X = (X_0, \ldots, X_K)$	vector of risk factors
$\alpha_k = 1/\xi_k$	shape parameter of X_k
$\beta_k = \xi_k$	scale parameter of X_k
$w_{k,n}$	factor loading, measuring the sensitivity of obligor n to the risk factor X_k

Part I

Introduction to Credit Risk Modeling

1
Risk Measurement

These lecture notes focus on the measurement and management of concentration risk in credit portfolios. In the first part of these lecture notes we will give a brief introduction to credit risk modeling. We start with a review of well-known risk measures in Chapter 2, which we will frequently use in Parts II and III. In Chapter 3 we present the famous Merton model in some detail as most of the approaches in Part II are based on this framework. In Chapter 4 we presents the theoretical model underlying the Basel II risk weight functions, namely the Asymptotic Single Risk Factor model. Chapter 5 is devoted to the presentation of mixture models. As an example of this class of credit risk models, we introduce in Chapter 6 the CreditRisk$^+$ model.

Let us start with a definition of what is meant by credit risk. *Credit risk* or *credit worthiness* is the risk of loss due to a counterparty defaulting on a contract or, more generally, the risk of loss due to some "credit event". In the ECB glossary credit risk is defined as "the risk that a counterparty will not settle an obligation in full, either when due or at any time thereafter." Traditionally this is applied to bonds where debt holders were concerned that the counterparty, to whom they've made a loan, might default on a payment (coupon or principal). For that reason, credit risk is sometimes also called *default risk*. These definitions indicate that credit risk is closely related to randomness. Hence, in credit risk modeling one tries to make probabilistic assessments of the likelihood of default.

The uncertainty about future events depends on several variables of risk which we will discuss in Section 1.1. For a more detailed discussion of these risk drivers we refer to [21], Section 1. In Section 1.2 we will present the general framework for modeling credit portfolio risk and which represents the basic setting for most of the models in these lecture notes. Section 1.3 then presents a simplified framework of portfolio credit risk models where all obligors in the considered portfolio are in some sense exchangeable.

1.1 Variables of Risk

The main difficulties when modeling credit risk arise from the fact that default events are quite rare and that they occur unexpectedly. When default events take place, however, they often lead to significant losses, the size of which is not known before default. Although default events occur very seldom, credit risk is by definition inherent in almost everything that represents a payment obligation. This is due to the fact that the probability, that a certain obligor might default, can be very small but it is seldom 0. Hence even very high rated borrowers have a potential to default on their financial obligations. The uncertainty whether an obligor will default or not, also termed *arrival risk*, is measured by its *probability of default* (PD). For comparison reasons it is usually specified with respect to a given time horizon, typically one year. Then the probability of default describes the probability that the default event occurs before the specified time horizon.

Assigning a default probability to every borrower in the bank's credit portfolio, however, is quite complicated. One approach is to calibrate default probabilities from market data. This is done, for example, in the concept of Expected Default Frequencies (EDF) from KMV Corporation. Another method is to compute the default probabilities from the credit risk inherent in certain credit spreads of traded products, e.g. credit derivatives such as *credit default swaps* (CDS). Default probabilities can also be calibrated from ratings that are assigned to borrowers either by external rating agencies such as *Moody's Investor Services, Standard & Poor's* or *Fitch*, or by bank-internal rating methods. Closely related to the default probability is also the notion of *migration risk*, i.e. the risk of losses due to changes in the default probability.

The *exposure at default* EAD of an obligor denotes the portion of the exposure to the obligor which is lost in case of default.

If an obligor defaults it does not necessarily mean that the creditor receives nothing from the obligor. There is a chance that the obligor will partly recover, meaning that the creditor might receive an uncertain fraction of the notional value of the claim. The *recovery risk* describes this uncertainty about the severity of the losses if a default has happened. In the event of recovery, the uncertain quantity is the actual payoff that a creditor receives after a default. The probability distribution of the recovery rate, i.e. the probability that the recovery rate is of a given magnitude conditional upon default, then describes the recovery risk.

The *loss given default* (LGD) of a transaction describes the extent of the loss incurred in the event of default. Hence it is closely linked to the recovery rate by "1 minus recovery rate". It is usually modeled as a random variable describing the severity of losses in the default event. In case the applied credit risk model admits only a constant value for LGD, one usually chooses the expectation of this severity. Typical values lie between 45% and 80%.

Default correlation describes the degree to which the default risk of one firm depends on the default risk of another firm. Therefore, *default correlation risk* represents the risk that several obligors default together. It is characterized by the joint default probabilities of the obligors over a given time horizon.

Default dependence has a strong impact on the upper tail of a credit loss distribution for a large portfolio. There are a lot of economic reasons for expecting default dependencies. The financial health of a firm, for example, varies with randomly fluctuating macroeconomic factors, such as changes in economic growth. Since different firms are effected by common macroeconomic factors, we have dependence between their defaults. Moreover there might exist some direct economic links between firms, such as a strong borrower-lender relationship, which can also lead to default dependencies. The topic of default dependencies will also play an important role when dealing with concentration of risk and default contagion in credit portfolios. We will discuss this topic in more detail later on in Parts II and III.

In general, credit risk can be split in *expected losses* (EL), which can be forecasted and thus can easily be managed, and *unexpected losses* (UL), which are more complicated to quantify. The latter can again be distinguished in *systematic risk* and *idiosyncratic risk*. The former arises from dependencies across individual obligors in the portfolio and from common shocks, while the latter arises from obligor specific shocks. Idiosyncratic risk can be diversified away, whereas systematic risk cannot be eliminated but can be reduced by shifting exposures in a way to reduce correlation. *Economic capital* (EC) is held for unexpected losses that arise from systematic and idiosyncratic risk. We will provide a more rigorous definition of these risk measures in the next chapter.

1.2 The General Model Setting

To represent the uncertainty about future events, we specify a probability space $(\Omega, \mathcal{F}, \mathbb{P})$ with sample space Ω, σ-algebra $\mathcal{F}$, probability measure $\mathbb{P}$ and with filtration $(\mathcal{F}_t)_{t \geq 0}$ satisfying the usual conditions. We fix a time horizon $T > 0$. Usually T will equal one year.

As these lecture notes focus on the quantification of concentration risk, which can only be measured on portfolio level, we specify, as a general setup for the later chapters, a portfolio consisting of N risky loans indexed by $n = 1, \ldots, N$. We assume that the exposures have been aggregated so that there is a unique obligor for each position. Denote by EAD_n the exposure at default of obligor n and by

$$s_n = \mathrm{EAD}_n / \sum_{k=1}^{N} \mathrm{EAD}_k$$

its share on the total portfolio exposure. Especially, we have $\sum_{n=1}^{N} s_n = 1$. Suppose that all loans are ordered in decreasing exposure size, unless otherwise stated. We denote by PD_n the (one year) default probability of obligor n. Let LGD_n be the *loss given default* of obligor n and denote its expectation and volatility by ELGD_n and VLGD_n, respectively.

Denote by D_n the *default indicator* of obligor n, in the time period $[0, T]$. At time $t = 0$, all obligors are assumed to be in a non-default state. Since obligor n can either default or survive, D_n can be represented as a Bernoulli random variable taking the values

$$D_n = \begin{cases} 1 & \text{if obligor n defaults} \\ 0 & \text{otherwise} \end{cases}$$

with probabilities $\mathbb{P}(D_n = 1) = \text{PD}_n$ and $\mathbb{P}(D_n = 0) = 1 - \text{PD}_n$. Let $D = (D_1, \ldots, D_N)$ denote the corresponding binary random vector of default indicators. The joint default probability function is given by

$$p(d) = \mathbb{P}(D_1 = d_1, \ldots, D_N = d_N)$$

for $d \in \{0, 1\}^N$ and the marginal default probabilities are $\text{PD}_n = \mathbb{P}(D_n = 1)$ for $n = 1, \ldots, N$. Furthermore, the total number of defaulted obligors at time T is given by the random variable

$$M := \sum_{n=1}^{N} D_n.$$

When we discuss models for credit migrations it is convenient to define a random variable indicating the current state of an obligor, as for example the current rating grade. Therefore, we define the *state indicator* variable S_n of obligor n at time T as a random variable taking integer values in the set $\{0, 1, \ldots, R\}$ where R is the number of states or rating classes. The value 0 is interpreted as default, i.e. $D_n = 1$ corresponds to $S_n = 0$. Define $S = (S_1, \ldots, S_N)$ as the corresponding random vector of state indicators.

The *portfolio loss* L_N is defined as the random variable

$$L_N = \sum_{n=1}^{N} \text{EAD}_n \cdot \text{LGD}_n \cdot D_n. \tag{1.1}$$

The subscript N in L_N indicates the number of obligors in the portfolio. In most cases we will omit this index when there is no ambiguity about the portfolio size. We will sometimes also use the notation L_n to denote the loss variable of obligor n. In these cases, however, the number N of obligors in the whole portfolio will be fix so that there should be no confusion about what is meant by the index n. Unless otherwise stated we will always suppose that the following assumption holds.

Assumption 1.2.1
The exposure at default EAD_n, *the loss given default variable* LGD_n *and the default indicator* D_n *of any obligor n are independent.*

Of course, the default indicators of different obligors n and m can be correlated and, in any sensible model, also will be correlated leading to the important topic of default correlations. We will extensively discuss various models to account for default correlations in Parts II and III.

1.3 Exchangeable Models

In some situations we will simplify the general framework of Section 1.2 by focusing only on so-called exchangeable models. Therefore, recall the following definition from [61].

Definition 1.3.1 (Exchangeability)
A random vector S is said to be *exchangeable* if for any permutation Π the following equality in distribution holds

$$(S_1, \ldots, S_N) \stackrel{d}{=} (S_{\Pi(1)}, \ldots, S_{\Pi(N)}).$$

Remark 1.3.2 Note that if $S_1, \ldots, S_N$ are independent and identically distributed then they are exchangeable but not vice versa.

Hence, an exchangeable portfolio model is characterized by an exchangeable state indicator vector meaning that, for a selected subgroup of $k \in \{1, \ldots, N-1\}$ obligors, all $\binom{N}{k}$ possible k-dimensional marginal distributions of S (and, thus, of the default indicator D) are identical. Denote by

$$\pi_k := \mathbb{P}(D_{n_1} = 1, \ldots, D_{n_k} = 1)$$

for $\{n_1, \ldots, n_k\} \subset \{1, \ldots, N\}$ and $1 \leq k \leq N$, the joint default probability of an arbitrary subgroup of k obligors and by $\pi := \pi_1$ the default probability of a single obligor in the portfolio. In case of an exchangeable model we can then express the correlation between two different default indicators in terms of the first and second order joint default probabilities π and π_2

$$\mathrm{Corr}(D_n, D_m) = \varrho_D = \frac{\mathrm{Cov}(D_n, D_m)}{\mathbb{E}[D_n^2] - \mathbb{E}[D_n]^2} = \frac{\pi_2 - \pi^2}{\pi - \pi^2},$$

since one can easily check that $\mathbb{E}[D_n] = \pi$ and $\mathbb{E}[D_n D_m] = \pi_2$.[1]

---
[1] See [103], p. 345, for details.

2
Modeling Credit Risk

In this chapter we present some simple approaches to measure credit risk. We start in Section 2.1 with a short overview of the *standardized approach* of the Basel framework for banking supervision. This approach is a representative of the so-called *notional-amount approach*. In this concept, the risk of a portfolio is defined as the sum of the notional values of the individual securities in the portfolio, where each notional value may be weighted by a certain risk factor, representing the riskiness of the asset class to which the security belongs. The advantage of this approach is its apparent simplicity, however, it has several drawbacks as, for example, netting and diversification effects are not taken into account.

One main challenge of credit risk management is to make default risks assessable. For this purpose we present several *risk measures* based on the portfolio loss distributions. These are typically statistical quantities describing the conditional or unconditional loss distribution of the portfolio over some predetermined time horizon. The *expected* and *unexpected loss*, which we present in Section 2.2, are defined as the expectation and standard deviation, respectively, of the portfolio loss variable. Hence, they belong to this class of risk measures. Further representatives are the *Value-at-Risk* (VaR) and the *Expected Shortfall* (ES) which we discuss in Sections 2.3 and 2.4. Based on the expected loss and Value-at-Risk we introduce in Section 2.5 the concept of *economic capital* of a portfolio. All of these risk measures have a lot of advantages as, for example, the aggregation from a single position to the whole portfolio makes sense in this framework. Moreover, diversification effects and netting can be reflected and the loss distributions are comparable across portfolios. However, the problem is that any estimate of the loss distribution is based on past data which are of limited use in predicting future risk. Furthermore, it is in general difficult to estimate the loss distribution accurately, particularly for large portfolios. Models that try to predict the future development of the portfolio loss variable will the studied in later chapters.

2.1 The Regulatory Framework

The First Basel Accord of 1988, also known as Basel I, laid the basis for international minimum capital standard and banks became subject to *regulatory capital requirements,* coordinated by the Basel Committee on Banking Supervision. This committee has been founded by the Central Bank Governors of the Group of Ten (G10) at the end of 1974.

The cause for Basel I was that, in the view of the Central Bank Governors of the Group of Ten, the equity of the most important internationally active banks decreased to a worrisome level. The downfall of Herstatt-Bank underpinned this concern. Equity is used to absorb losses and to assure liquidity. To decrease insolvency risk of banks and to minimize potential costs in the case of a bankruptcy, the target of Basel I was to assure a suitable amount of equity and to create consistent international competitive conditions.

The rules of the Basel Committee do not have any legal force. The supervisory rules are rather intended to provide guidelines for the supervisory authorities of the individual nations such that they can implement them in a suitable way for their banking system.

The main focus of the first Basel Accord was on credit risk as the most important risk in the banking industry. Within Basel I banks are supposed to keep at least 8% equity in relation to their assets. The assets are weighted according to their degree of riskiness where the risk weights are determined for four different borrower categories shown in Table 2.1.

Table 2.1. Risk weights for different borrower categories

Risk Weight in %	0	10	50	100
Borrower Category	State	Bank	Mortgages	Companies and Retail Customers

The required equity can then be computed as

$$\text{Minimal Capital} = \text{Risk Weighted Assets} \times 8\%.$$

Hence the portfolio credit risk is measured as the sum of *risk weighted assets,* that is the sum of notional exposures weighted by a coefficient reflecting the creditworthiness of the counterparty (the risk weight).

Since this approach did not take care of market risk, in 1996 an amendment to Basel I has been released which allows for both a *standardized approach* and a method based on internal Value-at-Risk (VaR) models for *market risk* in larger banks. The main criticism of Basel I, however, remained. Namely, it does not account for methods to decrease risk as, for example, by means of portfolio diversification. Moreover, the approach measures risk in an insufficiently differentiated way since minimal capital requirements are computed independent of the borrower's creditworthiness. These drawbacks lead to the development of the Second Basel Accord from 2001 onwards. In June 2004 the Basel Committee on Banking Supervision released a *Revised Framework on International convergence of capital measurement and capital standards* (in short: Revised Framework or Basel II). The rules officially came into force on January 1st, 2008, in the European Union. However, in practice they have been applied already before that date. The main targets of Basel II are the same as in Basel I as well. However, Basel II focuses not only on *market* and *credit risk* but also puts *operational risk* on the agenda.

Basel II is structured in a *three-pillar framework*. Pillar 1 sets out details for adopting more risk sensitive minimal capital requirements, so-called *regulatory capital,* for banking organizations, Pillar 2 lays out principles for the *supervisory review process* of capital adequacy and Pillar 3 seeks to establish *market discipline* by enhancing transparency in banks' financial reporting.

The former regulation lead banks to reject riskless positions, such as asset-backed transactions, since risk weighted assets for these positions were the same as for more risky and more profitable positions. The main goal of Pillar 1 is to take care of the specific risk of a bank when measuring minimal capital requirements. Pillar 1 therefore accounts for all three types of risk: credit risk, market risk and operational risk.

Concerning credit risk the new accord is more flexible and risk sensitive than the former Basel I accord. Within Basel II banks may opt for the *standard approach* which is quite conservative with respect to capital charge and the more advanced, so-called *Internal Ratings Based* (IRB) approach when calculating regulatory capital for credit risk. In the standard approach, credit risk is measured by means of external ratings provided by certain rating agencies such as Standard&Poor's, Moody's or Fitch Ratings. In the IRB approach, the bank evaluates the risk itself. This approach, however, can only be applied when the supervisory authorities accept it. The bank, therefore, has to prove that certain conditions concerning the method and transparency are fulfilled. Basel II distinguishes between expected loss and unexpected loss. The former directly charges equity whereas for the latter banks have to keep the appropriate capital requirements.

The capital charge for market risk within Basel II is similar to the approach in the amendment of 1996 for Basel I. It is based mainly on VaR approaches that statistically measure the total amount a bank can maximally lose.

A basic innovation of Basel II was the creation of a new risk category, operational risk, which is explicitly taken into account in the new accord.

The supervisory review process of Pillar 2 is achieved by the supervisory authorities which evaluate and audit the compliance of regulations with respect to methods and transparency which are necessary for a bank to be allowed to use internal ratings.

The main target of Pillar 3 is to improve market discipline by means of transparency of information concerning a bank's external accounting. Transparency can, for example, increase the probability of a decline in a bank's own stocks and therefore, motivate the bank to hold appropriate capital for potential losses.

2.2 Expected and Unexpected Loss

Although it is in general not possible to forecast the losses, a bank will suffer in a certain time period, a bank can still predict the average level of credit loss, it can expect to experience for a given portfolio. These losses are referred to as the *expected loss* (EL) and are simply given by the expectation of the portfolio loss variable L defined by equation (1.1). Note that we omit the index N here as the number N of obligors is fixed in this chapter. We will use the index n to refer to quantities specific to obligor n. The expected loss EL_n on a certain obligor n represents a kind of *risk premium* which a bank can charge for taking the risk that obligor n might default. It is defined as

$$\mathrm{EL}_n = \mathbb{E}[L_n] = \mathrm{EAD}_n \cdot \mathrm{ELGD}_n \cdot \mathrm{PD}_n,$$

since the expectation of any Bernoulli random variable is its event probability. The *expected loss reserve* is the collection of risk premiums for all loans in a given credit portfolio. It is defined as the expectation of the portfolio loss L and, by additivity of the expectation operator, it can be expressed as

$$\mathrm{EL} = \sum_{n=1}^{N} \mathrm{EAD}_n \cdot \mathrm{ELGD}_n \cdot \mathrm{PD}_n.$$

As one of the main reasons for banks holding capital is to create a protection against peak losses that exceed expected levels, holding only the expected loss reserve might not be appropriate. Peak losses, although occurring quite seldom, can be very large when they occur. Therefore, a bank should also reserve money for so-called *unexpected losses* (UL). The deviation of losses from the EL is usually measured by means of the standard deviation of the loss variable. Therefore, the unexpected loss UL_n on obligor n is defined as

$$\mathrm{UL}_n = \sqrt{\mathbb{V}[L_n]} = \sqrt{\mathbb{V}[\mathrm{EAD}_n \cdot \mathrm{LGD}_n \cdot D_n]}.$$

In case the default indicator D_n, and the LGD variable are uncorrelated (and the EAD is constant), the UL on borrower n is given by

$$\mathrm{UL}_n = \mathrm{EAD}_n \sqrt{\mathrm{VLGD}_n^2 \cdot \mathrm{PD}_n + \mathrm{ELGD}_n^2 \cdot \mathrm{PD}_n(1 - \mathrm{PD}_n)},$$

where we used that for Bernoulli random variables D_n the variance is given by $\mathbb{V}[D_n] = \mathrm{PD}_n \cdot (1 - \mathrm{PD}_n)$.

On the portfolio level, additivity holds for the variance UL^2 if the default indicator variables of the obligors in the portfolio are pairwise uncorrelated; due to Bienaymé's Theorem. If they are correlated, additivity is lost. Unfortunately this is the standard case and leads to the important topic of correlation modeling with which we will deal later on. In the correlated case, the unexpected loss of the total portfolio is given by

$$\mathrm{UL} = \sqrt{\mathbb{V}[L]} = \sqrt{\sum_{n=1}^{N}\sum_{k=1}^{N} \mathrm{EAD}_n \cdot \mathrm{EAD}_k \cdot \mathrm{Cov}\left[\mathrm{LGD}_n \cdot D_n; \mathrm{LGD}_k \cdot D_k\right]}$$

and, for constant loss given defaults ELGD_n, this equals

$$\mathrm{UL}^2 = \sum_{n,k=1}^{N} \mathrm{EAD}_n\, \mathrm{EAD}_k\, \mathrm{ELGD}_n\, \mathrm{ELGD}_k\, \varrho_{n,k} \sqrt{\mathrm{PD}_n(1-\mathrm{PD}_n)\,\mathrm{PD}_k(1-\mathrm{PD}_k)}$$

where $\varrho_{n,k} \equiv \mathrm{Corr}[D_n, D_k]$.

2.3 Value-at-Risk

As the probably most widely used risk measure in financial institutions we will briefly discuss *Value-at-Risk* (VaR) in this section. Here and in the next section we mainly follow the derivations in [103], pp. 37-48, to which we also refer for more details.

Value-at-Risk describes the maximally possible loss which is not exceeded in a given time period with a given high probability, the so-called confidence level. A formal definition is the following.[1]

Definition 2.3.1 (Value-at-Risk)
Given some confidence level $q \in (0, 1)$. The *Value-at-Risk* (VaR) of a portfolio with loss variable L at the confidence level q is given by the smallest number x such that the probability that L exceeds x is not larger than $(1-q)$. Formally,

$$\mathrm{VaR}_q(L) = \inf\{x \in \mathbb{R} : \mathbb{P}(L > x) \leq 1 - q\} = \inf\{x \in \mathbb{R} : F_L(x) \geq q\}.$$

Here $F_L(x) = \mathbb{P}(L \leq x)$ is the distribution function of the loss variable.

Thus, VaR is simply a quantile of the loss distribution. In general, VaR can be derived for different holding periods and different confidence levels. In credit risk management, however, the holding period is typically one year and typical values for q are 95% or 99%. Today higher values for q are more

---
[1] Compare [103], Definition 2.10.

and more common. The confidence level q in the Second Basel Accord is e.g. 99.9% whereas in practice a lot of banks even use a 99.98% confidence level. The reason for these high values for q is that banks want to demonstrate external rating agencies a solvency level that corresponds at least to the achieved rating class. A higher confidence level (as well as a longer holding period) leads to a higher VaR.

We often use the alternative notation $\alpha_q(L) := \text{VaR}_q(L)$. If the distribution function F of the loss variable is continuous and strictly increasing, we simply have $\alpha_q(L) = F^{-1}(q)$, where F^{-1} is the ordinary inverse of F.

Example 2.3.2 Suppose the loss variable L is normally distributed with mean μ and variance σ^2. Fix some confidence level $q \in (0,1)$. Then

$$\text{VaR}_q(L) = \mu + \sigma \Phi^{-1}(q)$$

where Φ denotes the standard normal distribution function and $\Phi^{-1}(q)$ the qth quantile of Φ. To prove this, we only have to show that $F_L(\text{VaR}_q(L)) = q$ since F_L is strictly increasing. An easy computation shows the desired property

$$\mathbb{P}(L \leq \text{VaR}_q(L)) = \mathbb{P}\left(\frac{L-\mu}{\sigma} \leq \Phi^{-1}(q)\right) = \Phi\left(\Phi^{-1}(q)\right) = q.$$

Proposition 2.3.3 *For a deterministic monotonically decreasing function $g(x)$ and a standard normal random variable X the following relation holds*

$$\alpha_q(g(X)) = g(\alpha_{1-q}(X)) = g(\Phi^{-1}(1-q)).$$

Proof. Indeed, we have

$$\alpha_q(g(X)) = \inf\{x \in \mathbb{R} : \mathbb{P}(g(X) \geq x) \leq 1-q\}$$
$$= \inf\{x \in \mathbb{R} : \mathbb{P}(X \leq g^{-1}(x)) \leq 1-q\}$$
$$= \inf\{x \in \mathbb{R} : \Phi(g^{-1}(x)) \leq 1-q\}$$
$$= g(\Phi^{-1}(1-q)).$$

which proves the assertion. □

By its definition, however, VaR gives no information about the severity of losses which occur with a probability less than $1-q$. If the loss distribution is heavy tailed, this can be quite problematic. This is a major drawback of the concept as a risk measure and also the main intention behind the innovation of the alternative risk measure *Expected Shortfall* (ES) which we will present in the next section. Moreover, VaR is not a *coherent* risk measure since it is not subadditive (see [7], [8]). Non-subadditivity means that, if we have two loss distributions F_{L_1} and F_{L_2} for two portfolios and if we denote the overall loss distribution of the merged portfolio $L = L_1 + L_2$ by F_L, then we do not necessarily have that $\alpha_q(F_L) \leq \alpha_q(F_{L_1}) + \alpha_q(F_{L_2})$. Hence, the VaR of the merged portfolio is not necessarily bounded above by the sum of the VaRs of the individual portfolios which contradicts the intuition of diversification benefits associated with merging portfolios.

2.4 Expected Shortfall

Expected Shortfall (ES) is closely related to VaR. Instead of using a fixed confidence level, as in the concept of VaR, one averages VaR over all confidence levels $u \geq q$ for some $q \in (0,1)$. Thus, the tail behavior of the loss distribution is taken into account. Formally, we define ES as follows.[2]

Definition 2.4.1 (Expected Shortfall)
For a loss L with $\mathbb{E}[|L|] < \infty$ and distribution function F_L, the *Expected Shortfall* (ES) at confidence level $q \in (0,1)$ is defined as

$$\mathrm{ES}_q = \frac{1}{1-q} \int_q^1 \mathrm{VaR}_u(L) du.$$

By this definition it is obvious that $\mathrm{ES}_q \geq \mathrm{VaR}_q$. If the loss variable is integrable with continuous distribution function, the following Lemma holds.

Lemma 2.4.2 *For integrable loss variable L with continuous distribution function F_L and any $q \in (0,1)$, we have*

$$\mathrm{ES}_q = \frac{\mathbb{E}\left[L; L \geq \mathrm{VaR}_q(L)\right]}{1-q} = \mathbb{E}\left[L | L \geq \mathrm{VaR}_q(L)\right],$$

where we have used the notation $\mathbb{E}[X; A] \equiv \mathbb{E}[X \mathbf{1}_A]$ for a generic integrable random variable X and a generic set $A \in \mathcal{F}$.

For the proof see [103], page 45. Hence, in this situation expected shortfall can be interpreted as the expected loss that is incurred in the event that VaR is exceeded. In the discontinuous case, a more complicated formula holds

$$\mathrm{ES}_q = \frac{1}{1-q}\left(\mathbb{E}\left[L; L \geq \mathrm{VaR}_q(L)\right] + \mathrm{VaR}_q(L) \cdot (1 - q - \mathbb{P}(L \geq \mathrm{VaR}_q(L)))\right).$$

For a proof see Proposition 3.2 of [1].

Example 2.4.3 Suppose the loss distribution F_L is normal with mean μ and variance σ^2. Fix a confidence level $q \in (0,1)$. Then

$$\mathrm{ES}_q = \mu + \sigma \frac{\phi(\Phi^{-1}(q))}{1-q},$$

---

[2] Compare [103], Definition 2.15.

Fig. 2.1. VaR and ES for standard normal distribution

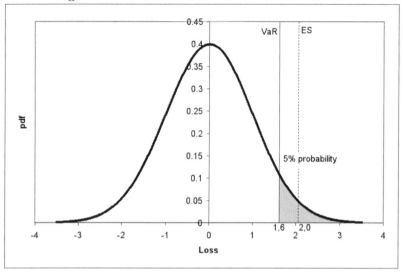

where ϕ is the density of the standard normal distribution. For the proof, note that
$$\text{ES}_q = \mu + \sigma \mathbb{E}\left[\frac{L-\mu}{\sigma}\bigg|\frac{L-\mu}{\sigma} \geq \alpha_q\left(\frac{L-\mu}{\sigma}\right)\right].$$
Hence it is sufficient to compute the expected shortfall for the standard normal random variable $\tilde{L} := (L-\mu)/\sigma$. Here we obtain
$$\text{ES}_q(\tilde{L}) = \frac{1}{1-q}\int_{\Phi^{-1}(q)}^{\infty} l\phi(l)dl = \frac{1}{1-q}[-\phi(l)]_{l=\Phi^{-1}(q)}^{\infty} = \frac{\phi(\Phi^{-1}(q))}{1-q}.$$

Figure 2.1 shows the probability density function of a standard normal random variable. The solid vertical line shows the Value-at-Risk at level 95% which equals 1.6, while the dashed vertical line indicates the Expected Shortfall at level 95% which is equal to 2.0. Hence, the grey area under the distribution function is the amount which will be lost with 5% probability.

For an example to demonstrate the sensitivity to the severity of losses exceeding VaR and its importance see [103], Example 2.2.1, pp. 46–47. In particular for heavy-tailed distributions, the difference between ES and VaR is more pronounced than for normal distributions. Figure 2.2 shows the probability density function of a $\Gamma(3,1)$ distributed random variable with vertical lines at its 95% Value-at-Risk and Expected Shortfall. The grey area under the distribution function is the portion which is lost with 5% probability. In this case, the Value-at-Risk at level 95% equals 6.3 while the Expected Short-

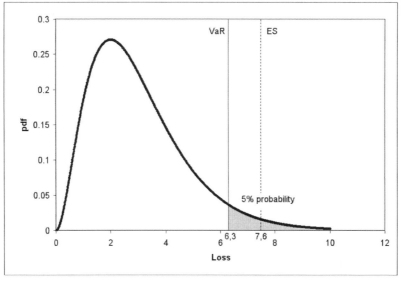

Fig. 2.2. VaR and ES for Gamma distribution

fall at level 95% for the $\Gamma(3,1)$ distribution equals 7.6.

Figures 2.1 and 2.2 also show that the ES for a distribution is always higher than the Value-at-Risk, a result we already derived theoretically in the above discussion.

2.5 Economic Capital

Since there is a significant likelihood that losses will exceed the portfolio's EL by more than one standard deviation of the portfolio loss, holding the UL of a portfolio as a risk capital for cases of financial distress might not be appropriate. The concept of *economic capital* (EC) is a widely used approach for bank internal credit risk models.

Definition 2.5.1 (Economic Capital)
The *economic capital* EC_q for a given confidence level q is defined as the Value-at-Risk $\alpha_q(L)$ at level q of the portfolio loss L minus the expected loss EL of the portfolio,
$$EC_q = \alpha_q(L) - EL.$$

For a confidence level $q = 99.98\%$, the EC_q can be interpreted as the (on average) appropriate capital to cover unexpected losses in $9,998$ out of $10,000$ years, where a time horizon of one year is assumed. Hence it represents the capital, a bank should reserve to limit the probability of default to a given confidence level. The VaR is reduced by the EL due to the common decomposition of total risk capital, that is VaR, into a part covering expected losses and a part reserved for unexpected losses.

Suppose a bank wants to include a new loan in its portfolio and, thus, has to adopt its risk measurement. While the EL is independent from the composition of the reference portfolio, the EC strongly depends on the composition of the portfolio in which the new loan will be included. The EC charge for a new loan of an already well diversified portfolio, for example, might be much lower than the EC charge of the same loan when included in a portfolio where the new loan induces some concentration risk. For this reason the EL charges are said to be *portfolio independent*, while the EC charges are *portfolio dependent* which makes the calculation of the contributory EC much more complicated, since the EC always has to be computed based on the decomposition of the complete reference portfolio.

In the worst case, a bank could lose its entire credit portfolio in a given year. Holding capital against such an unlikely event is economically inefficient. As banks want to spend most of their capital for profitable investments, there is a strong incentive to minimize the capital a bank holds. Hence the problem of risk management in a financial institution is to find the balance between holding enough capital to be able to meet all debt obligations also in times of financial distress, on the one hand, and minimizing economic capital to make profits, on the other hand.

3
The Merton Model

Credit risk models can be divided into two fundamental classes of models, *structural* or *asset-value models*, on the one hand, and *reduced-form* or *default-rate models*, on the other hand.

Asset-value models trace back to the famous Merton model [105], where the default of a firm is modeled in terms of the relationship between its assets and the liabilities that it faces at the end of a given time period. The value of the firm's debt at maturity then equals the nominal value of the liabilities minus the pay-off of a European put option on the firm's value. The asset value process is modeled as a geometric Brownian motion and default occurs when the asset value at maturity is lower than the liabilities. In this case the debtor hands over the firm and thus exercises the put option. The assumptions of the Merton model are in line with the Black-Scholes model (see [20]). Hence the risky credit can be priced with the Black-Scholes option pricing theory. The Merton model has been developed further by several authors to include also the possibility of bankruptcy before maturity or the stochastic evolution of the risk free rate, to name just two. In models of this type default risk depends mainly on the stochastic evolution of the asset value and default occurs when the random variable describing this asset value falls below a certain threshold which represents the liabilities. Therefore these structural models are also referred to as *threshold* or *latent variable models*. A general definition of such a latent variable model is the following.[1]

Definition 3.0.2 (Latent variable model)
Let $V = (V_1, ..., V_N)$ be an N-dimensional vector with continuous marginal distribution functions $F_n(v) = \mathbb{P}(V_n \leq v)$. Given a sequence of deterministic thresholds

$$-\infty = b_0^n < b_1^n < ... < b_R^n < b_{R+1}^n = \infty$$

we say that obligor n is in state $S_n = r$ if and only if $V_n \in (b_r^n, b_{r+1}^n]$ for some $r \in \{0, 1, ..., R\}$ and $n \in \{1, ..., N\}$. Then $(V_n, (b_r^n)_{0 \leq r \leq R+1})_{1 \leq n \leq N}$ defines

---
[1] Compare [61], Definition 3.1.

a *latent variable model* for the state vector $S = (S_1, \ldots, S_N)$. The individual default probability of firm n is given by $F_n(b_1^n) = \mathbb{P}(V_n \leq b_1^n)$.

By determination of the increasing sequence of thresholds, the migration over non default grades is captured and can be interpreted as rating classes. Main industry representatives of this class of credit risk models are the KMV model and the CreditMetrics model which we discuss in Subsections 3.3.1 and 3.3.2 below.

In *reduced-form models* one directly models the process of credit defaults instead of constructing a stochastic process of the firm's asset value which indirectly leads to a model of the firm's default. In this class of models, defaults or rating migrations can occur in any discrete time interval and only the probability of default is specified. The default probability of a firm is usually modeled as a non-negative random variable, whose distribution typically depends on economic covariables. Therefore this class of models is sometimes also called *mixture models*. We will give a brief introduction to this class of models in Chapter 5. Industry examples for this class of credit risk models are CreditRisk$^+$ and Credit Portfolio View. We will discuss CreditRisk$^+$ in more detail in Chapter 6.

3.1 The General Framework

The model proposed by [105] is the precursor of all asset-value models and, although a lot of extensions have been developed since that time, the original Merton model remains influential and is still popular in the practice of credit risk analysis. As some of the models, we will study in Parts II and III, are based on the Merton model, we will discuss the latter here in some detail.

The Merton model assumes the asset value of a firm to follow some stochastic process $(V_t)_{t \geq 0}$. There are only two classes of securities; equity and debt. It is assumed that equity receives no dividends and that the firm cannot issue new debt. The model assumes that the company's debt is given by a zero-coupon bond with face value B that will become due at a future time T. The firm defaults if the value of its assets is less than the promised debt repayment at time T. In the Merton model default can occur only at the maturity T of the bond. Denote the value at time t of equity and debt by S_t and B_t, respectively. In a frictionless market (i.e. when there are no taxes or transaction costs), the value of the firm's assets is given by the sum of debt and equity, i.e.

$$V_t = S_t + B_t, \ 0 \leq t \leq T.$$

At maturity there are only two possible scenarios:

(i) $V_T > B$: the value of the firm's assets exceeds the debt. In this case the debtholders receive $B_T = B$, the shareholders receive the residual value $S_T = V_T - B$, and there is no default.

(ii) $V_T \leq B$: the value of the firm's assets is less than its debt. Hence the firm cannot meet its financial obligations and defaults. In this case, the debtholders take ownership of the firm, and the shareholders are left with nothing, so that we have $B_T = V_T$, $S_T = 0$.

Combining the above two results, the payment to the shareholders at time T is given by
$$S_T = \max(V_T - B, 0) = (V_T - B)^+ \tag{3.1}$$
and debtholders receive
$$B_T = \min(V_T, B) = B - (B - V_T)^+. \tag{3.2}$$

This shows that the value of the firm's equity is the payoff of a European call option on the assets of the firm with strike price equal to the promised debt payment. By put-call parity, the firm's debt comprises a risk-free bond that guarantees payment of B plus a short European put option on the firm's assets with exercise price equal to the promised debt payment B. The Merton model thus treats the asset value V_t as any underlying. It assumes that under the real-world probability measure $\mathbb{P}$ the asset value process $(V_t)_{t \geq 0}$ follows a geometric Brownian motion of the form
$$dV_t = \mu_V V_t dt + \sigma_V V_t dW_t, \quad 0 \leq t \leq T, \tag{3.3}$$
for constants $\mu_V \in \mathbb{R}$, $\sigma_V > 0$, and a standard Brownian motion $(W_t)_{t \geq 0}$. Further, it makes all other simplifying assumptions of the Black-Scholes option pricing formula (see [20]). The solution at time T of the stochastic differential equation (3.3) with initial value V_0 can be computed and is given by
$$V_T = V_0 \exp\left(\left(\mu_V - \frac{1}{2}\sigma_V^2\right)T + \sigma_V W_T\right).$$

This implies, in particular, that
$$\ln V_T \sim \mathcal{N}\left(\ln V_0 + \left(\mu_V - \frac{1}{2}\sigma_V^2\right)T, \sigma_V^2 T\right).$$

Hence the market value of the firm's equity at maturity T can be determined as the price of a European call option on the asset value V_t with exercise price B and maturity T. The risk neutral pricing theory then yields that the market value of equity at time $t < T$ can be computed as the discounted expectation (under the risk-neutral equivalent measure $\mathbb{Q}$) of the payoff function (3.1), i.e.
$$S_t = \mathbb{E}^{\mathbb{Q}}\left[e^{-r(T-t)}(V_T - B)^+ \Big| \mathcal{F}_t\right],$$
and is given by
$$S_t = V_t \cdot \Phi(d_{t,1}) - B \cdot e^{-r(T-t)} \cdot \Phi(d_{t,2}),$$

where

$$d_{t,1} = \frac{\ln(V_t/B) + (r + \frac{1}{2}\sigma_V^2)(T-t)}{\sigma_V\sqrt{T-t}} \quad \text{and} \quad d_{t,2} = d_{t,1} - \sigma_V\sqrt{T-t}.$$

Here r denotes the risk-free interest rate which is assumed to be constant. According to equation (3.2) we can value the firm's debt at time $t \leq T$ as

$$B_t = \mathbb{E}^{\mathbb{Q}}\left[e^{-r(T-t)}\left(B - (B - V_T)^+\right)\Big|\mathcal{F}_t\right]$$
$$= Be^{-r(T-t)} - \left(Be^{-r(T-t)}\Phi(-d_{t,2}) - V_t\Phi(-d_{t,1})\right).$$

We discount the payment B at the risk-free rate because that payment is risk-free since we stripped out the credit risk as a put option.

The default probability of the firm by time T is the probability that the shareholders will not exercise their call option to buy the assets of the company for B at time T, i.e. it is precisely the probability of the call option expiring *out-of-the-money*.[2] It can be computed as

$$\mathbb{P}(V_T \leq B) = \mathbb{P}(\ln V_T \leq \ln B) = \Phi\left(\frac{\ln(B/V_0) - (\mu_V - \frac{1}{2}\sigma_V^2)T}{\sigma_V\sqrt{T}}\right). \quad (3.4)$$

Equation (3.4) shows that the default probability is increasing in B, decreasing in V_0 and μ_V and, for $V_0 > B$, increasing in σ_V, which is all perfectly in line with economic intuition. Under the risk-neutral measure $\mathbb{Q}$ we have

$$\mathbb{Q}(V_T \leq B) = \mathbb{Q}\left(\frac{\ln(B/V_0) - (r - \frac{1}{2}\sigma_V^2)T}{\sigma_V\sqrt{T}} \leq -d_{0,2}\right) = 1 - \Phi(d_{0,2}).$$

Hence the risk-neutral default probability, given information up to time t, is given by $1 - \Phi(d_{t,2})$.

Remark 3.1.1 The Merton model can also incorporate credit migrations and, thus, is not limited to the default-only mode as presented above. Therefore, we consider a firm which has been assigned to some rating category at time $t_0 = 0$. The time horizon is fixed to $T > 0$. Assume that the transition probabilities $p(r)$ for a firm are available for all rating grades $0 \leq r \leq R$. The transition probability $p(r)$ thus denotes the probability that the firm belongs to rating class r at the time horizon T. In particular, $p(0)$ denotes the default probability PD of the firm.

Suppose that the asset-value process V_t of the firm follows the model given in (3.3). Define thresholds

---

[2] An option is said to be *in-the-money* if it has positive intrinsic value, respectively *out-of-the-money* if it has zero intrinsic value. A call is in-the-money if the value of the underlying is above the strike price. A put is in-the-money if the value of the underlying is below the strike price.

$$-\infty = b_0 < b_1 < \ldots < b_R < b_{R+1} = \infty$$

such that $\mathbb{P}(b_r < V_T \leq b_{r+1}) = p(r)$ for $r \in \{0, \ldots, R\}$, i.e. the probability that the firm belongs to rating r at the time horizon T equals the probability that the firm's value at time T is between b_r and b_{r+1}. Hence we have translated the transition probabilities into a series of thresholds for an assumed asset-value process. b_1 denotes the default threshold, i.e. the value of the firm's liabilities B. The higher thresholds are the asset-value levels marking the boundaries of higher rating categories.

Although the Merton model provides a useful context for modeling credit risk and, moreover, practical implementations of the model are used by many financial institutions, it also has some drawbacks. It assumes the firm's debt financing consists of a one-year zero coupon bond. For most firms, however, this is an oversimplification. Moreover, the simplifying assumptions of the Black-Scholes model are questionable in the context of corporate debt. In particular, the assumption of normally distributed losses can lead to an underestimation of the potential risk in a loan portfolio. Alternatively, [3] describes the portfolio loss within a one-factor Lévy model. Compared to a model with normally distributed asset returns, using a distribution with fatter tails as, for example, the Variance Gamma distribution, leads to an increase in the economic capital of the portfolio.

Finally, and this might be the most important shortcoming of the Merton model, the firm's value is not observable which makes assigning values to it and its volatility problematic.

3.2 The Multi-Factor Merton Model

In credit risk we wish to explain the firm's economic success by means of some global underlying influences. This aims at the derivation of so-called *factor models*. Factor models provide a possibility to interpret the correlation between single loss variables in terms of some underlying economic variables such that large portfolio losses can be explained by these economic factors. Moreover, factor models lead to a reduction of the computational effort which can also be controlled by the number of economic factors considered in the model. The Merton model can be understood as a multi-factor model as will be explained in this section. Here we focus solely on the default mode version of the Merton model. In the default-only mode, the Merton model is of Bernoulli type where the decision about default or survival of a firm at the end of a time period is made by comparing the firm's asset value to a certain threshold value. If the firm value is below this threshold, the firm defaults and otherwise it survives. This model is frequently used in the different approaches for measuring sector concentration which we will discuss in Chapter 10.

3 The Merton Model

We consider the portfolio of N borrowers described in Section 1.2. Each of the obligors has exactly one loan with principal EAD_n. We express the loan as a share of total portfolio exposure, i.e. the exposure share of obligor n is given by $s_n = \text{EAD}_n / \sum_{i=1}^{N} \text{EAD}_i$.

Fix a time horizon $T > 0$. The Merton model is a so-called asset-value model, meaning that the loss distribution is derived by focusing on a description of the firm's asset value. Therefore, we define $V_t^{(n)}$ to be the asset value of counterparty n at time $t \leq T$. For every counterparty there exists a threshold C_n such that counterparty n defaults in the time period $[0, T]$ if $V_T^{(n)} < C_n$, i.e. the asset value at maturity T is less than the threshold value. Think for example of C_n as counterparty n's liabilities. Hence $V_T^{(n)}$ can be viewed as a latent variable driving the default event. Therefore, we define for $n = 1, \ldots, N$,

$$D_n = \mathbf{1}_{\{V_T^{(n)} < C_n\}} \sim \text{Bern}\left(1; \mathbb{P}[V_T^{(n)} < C_n]\right) \tag{3.5}$$

where $\text{Bern}(1; p)$ denotes the Bernoulli distribution such that the event 1 occurs with probability p.

Let r_n denote borrower n's asset-value log-returns $\log\left(V_T^{(n)}/V_0^{(n)}\right)$. In the factor model approach, the main assumption is that the asset value process depends on some underlying factors which represent the industrial and regional influences as well as on some idiosyncratic term.

Assumption 3.2.1
The asset returns r_n depend linearly on K standard normally distributed risk factors $X = (X_1, \ldots, X_K)$ affecting borrowers' defaults in a systematic way as well as on a standard normally distributed idiosyncratic term ε_n. Moreover, ε_n is independent of the systematic factors X_k for every $k \in \{1, \ldots, K\}$ and the ε_n are uncorrelated.

Under this assumption, borrower n's asset value log-returns $\log\left(V_T^{(n)}/V_0^{(n)}\right)$, after standardization, admit a representation of the form[3]

$$r_n = \beta_n Y_n + \sqrt{1 - \beta_n^2}\, \varepsilon_n, \tag{3.6}$$

where Y_n denotes the firm's *composite factor* and ε_n represents the *idiosyncratic shock*. The factor loading β_n illustrates borrower n's sensitivity to the systematic risk. Hence, it captures the linear correlation between r_n and Y_n. Y_n can be decomposed into K independent factors $X = (X_1, \ldots, X_K)$ by

$$Y_n = \sum_{k=1}^{K} \alpha_{n,k} X_k. \tag{3.7}$$

---
[3] For a motivation of this representation in terms of equicorrelation models we refer to [103], Example 3.34, p. 104.

3.2 The Multi-Factor Merton Model

The weights $\alpha_{n,k}$ describe the dependence of obligor n on an industrial or regional sector k represented by factor X_k.

Since the idiosyncratic shocks and the risk factors are assumed to be independent, the correlation of the counterparties' asset returns depends only on the correlation of the composite factors Y_n. Computing the variances of both sides of equation (3.6) yields

$$\mathbb{V}[r_n] = \beta_n^2 \cdot \mathbb{V}[Y_n] + (1 - \beta_n^2) \cdot \mathbb{V}[\varepsilon_n].$$

This composition can be interpreted as splitting the total risk of counterparty n into a *systematic* and an *idiosyncratic* risk component. The quantity $\beta_n^2 \cdot \mathbb{V}[Y_n]$ then expresses how much of the volatility of r_n can be explained by the volatility of Y_n and, thus, quantifies the systematic risk of the counterparty n. The term $(1 - \beta_n^2) \cdot \mathbb{V}[\varepsilon_n]$ captures the idiosyncratic risk which cannot be explained by the common factors X_k. Since we assumed that the asset returns r_n, the systematic risk factors X_k and the idiosyncratic terms ε_n are all standard normally distributed, we have to make sure that Y_n has unit variance. Therefore, the coefficients $\alpha_{n,k}$ must satisfy $\sum_{k=1}^{K} \alpha_{n,k}^2 = 1$.

Now we can rewrite equation (3.5) as

$$D_n = \mathbf{1}_{\{r_n < c_n\}} \sim \text{Bern}\,(1; \mathbb{P}[r_n < c_n])$$

where c_n is the threshold corresponding to C_n after exchanging $V_T^{(n)}$ by r_n. Assume the time horizon to equal 1 year, i.e. $T = 1$. Denote the one-year default probability of obligor n, as usual, by PD_n. We have $\text{PD}_n = \mathbb{P}[r_n < c_n]$ and, since $r_n \sim \mathcal{N}(0,1)$, we obtain

$$c_n = \Phi^{-1}[\text{PD}_n],$$

where Φ denotes the cumulative standard normal distribution function and Φ^{-1} its inverse. Hence, in the factor representation, the condition $r_n < c_n$ can be written as

$$\varepsilon_n < \frac{\Phi^{-1}(\text{PD}_n) - \beta_n Y_n}{\sqrt{1 - \beta_n^2}}.$$

Hence, the one-year default probability of obligor n, conditional on the factor Y_n, is given by

$$\text{PD}_n(Y_n) = \Phi\left(\frac{\Phi^{-1}(\text{PD}_n) - \beta_n Y_n}{\sqrt{1 - \beta_n^2}}\right). \tag{3.8}$$

The only remaining random part is the factor Y_n. Representing Y_n by the independent systematic factors $X = (X_1, \ldots, X_K)$, the default probability of obligor n, conditional on a specification $x = (x_1, \ldots, x_K)$ of X, can be written as

3 The Merton Model

$$\mathrm{PD}_n(x) = \Phi\left(\frac{\Phi^{-1}(\mathrm{PD}_n) - \beta_n \sum_{k=1}^{K} \alpha_{n,k} x_k}{\sqrt{1-\beta_n^2}}\right).$$

Figure 3.1 shows the dependence of the conditional default probability on the state of the economy, i.e. on the systematic risk factor X, in a single factor setting. Here we used an unconditional default probability of 0.5% and an asset correlation of 20%. For a bad state of the economy, i.e. when the systematic risk factor takes negative values, the conditional default probability is higher than for better states of the economy when X takes positive values.

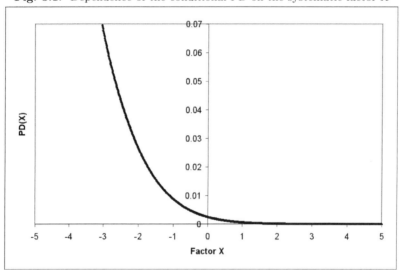

Fig. 3.1. Dependence of the conditional PD on the systematic factor X

Figure 3.2 shows the dependence of the conditional default probability on the unconditional default probability for a fixed correlation of 20% and three different stages of the economy. The dotted graph corresponds to a bad state of the economy where the systematic risk factor takes the value $X = -4$. The dashed graph corresponds to a risk factor $X = 0$ and the solid graph to a good state of the economy with $X = 4$.

We want to find an expression for the portfolio loss variable L.[4] If borrower n defaults, its rate of loss is determined by the variable loss given default, denoted LGD_n. Although a specific distribution is not assumed, we suppose

---

[4] Note that we skipped the index N here as the number N of obligors in our portfolio is fixed.

Fig. 3.2. Dependence of the conditional PD on the unconditional PD

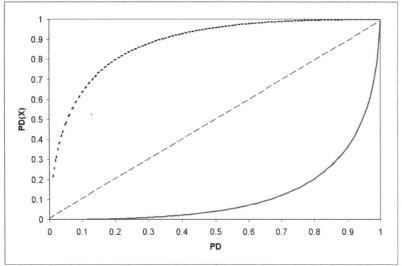

that LGD_n has mean $ELGD_n$ and standard deviation $VLGD_n$. Moreover, the loss given defaults are assumed to be independent for different borrowers as well as from all the other variables in the model. Thus, the portfolio loss rate can be written as

$$L = \sum_{n=1}^{N} s_n \, LGD_n \, \mathbf{1}_{\{r_n \leq \Phi^{-1}(PD_n)\}}. \tag{3.9}$$

Therefore, the expected loss rate of borrower n is the probability of default times the expected rate of loss in case of default $ELGD_n$. The expected portfolio loss rate consists of the exposure weighted sum of all expected individual losses. Since LGD_n is assumed to be independent from the default indicator $D_n := \mathbf{1}_{\{r_n \leq \Phi^{-1}(PD_n)\}}$ and, since the conditional expectation of the default indicator equals the probability that r_n lies below the threshold c_n conditional on the risk factors, we obtain

$$\mathbb{E}[L|(X_1, \ldots, X_K)] = \sum_{n=1}^{N} s_n \, ELGD_n \, \Phi\left(\frac{\Phi^{-1}(PD_n) - \beta_n \sum_{k=1}^{K} \alpha_{n,k} X_k}{\sqrt{1 - \beta_n^2}}\right). \tag{3.10}$$

The determination of the portfolio loss distribution requires a Monte Carlo simulation of the systematic risk factors $X_1, \ldots, X_K$. Let us explain this in more detail here. The default indicator variables D_n are Bernoulli distributed with parameter $PD_n = \mathbb{P}(D_n = 1)$ such that for some $d_n \in \{0, 1\}$ we have

$$\mathbb{P}(D_n = d_n) = PD_n^{d_n} \cdot (1 - PD_n)^{1-d_n}.$$

Now instead of the default probabilities PD_n, consider the conditional default probabilities $\mathrm{PD}_n(Y_n)$ given the composite factor Y_n. We can compute the joint distribution of the default indicator variables by integrating out the conditioning on the factors Y_n as

$$\mathbb{P}[D_1 = d_1, \ldots, D_N = d_N] = \int_{\mathbb{R}^N} \prod_{n=1}^{N} \mathrm{PD}_n(y)^{d_n} \cdot (1 - \mathrm{PD}_n(y))^{1-d_n} dF_Y(y),$$

for $d_n \in \{0, 1\}$, $n \in \{1, \ldots, N\}$, where we denoted the distribution function of the composite factors $Y_1, \ldots, Y_N$ by F_Y. Now we substitute $q_n = \mathrm{PD}_n(y)$, i.e. $y = \mathrm{PD}_n^{-1}(q_n)$. Note that the joint distribution function of the q_n's is given by the joint distribution function of the conditional PD_n's which are normally distributed. So we obtain

$$\mathbb{P}[D_1 = d_1, \ldots, D_N = d_N] = \int_{[0,1]^N} \prod_{n=1}^{N} q_n^{d_n} \cdot (1 - q_n)^{1-d_n} dF(q_1, \ldots, q_N),$$

where the distribution function F is explicitly given by

$$F(q_1, \ldots, q_N) = \Phi_N \left[\mathrm{PD}_1^{-1}(q_1), \ldots, \mathrm{PD}_N^{-1}(q_N); \Gamma \right],$$

where $\Phi_N[\cdot; \Gamma]$ denotes the cumulative multivariate centered Gaussian distribution with correlation matrix Γ and $\Gamma = (\varrho_{nm})_{1 \leq n,m \leq N}$ is the asset correlation matrix of the asset returns r_n.

Assuming a constant loss given default equal to ELGD_n for obligor n, the portfolio loss distribution can then be derived from the joint distribution of the default indicators as

$$\mathbb{P}[L \leq l] = \sum_{\substack{(d_1, \ldots, d_N) \in \{0,1\}^N; \\ \sum_{n=1}^{N} s_n \mathrm{ELGD}_n d_n \leq l}} \left(\sum_{n=1}^{N} s_n \mathrm{ELGD}_n d_n \right) \cdot \mathbb{P}[D_1 = d_1, \ldots, D_N = d_N].$$

Remark 3.2.2 In Section 10.1 we present an analytical approximation to receive the qth percentile of the loss distribution in this multi-factor framework under the assumption that portfolios are infinitely fine-grained such that the idiosyncratic risk is completely diversified away. This is one key assumption of the Asymptotic Single Risk Factor (ASRF) model which we will introduce in Chapter 4. Then Theorem 4.1.4 states that the portfolio loss converges almost surely to the expected loss conditional on the systematic risk factors,

$$L_N - \mathbb{E}[L_N | \{X_1, \ldots, X_K\}] \longrightarrow 0 \quad \text{almost surely},$$

as the number N of obligors in the portfolio increases to infinity and, thus, as idiosyncratic risk is diversified away.

3.3 Industry Models Based on the Merton Approach

In this section we will present two famous industry models descending from the Merton approach, namely the KMV model and the CreditMetrics model.

3.3.1 The KMV Model

As already mentioned, an important example of an industry model of the asset-value type is the KMV model, which was founded by KMV (a private company named after its founders Kealhofer, McQuown and Vasicek) in 1989 and which is now maintained by Moody's KMV. It uses the Merton approach in a slightly varied way to determine the risk of a credit portfolio. The main contribution of KMV, however, is not the theoretical model but its calibration to achieve that the default probabilities correspond to a large extend to the empirically observed ones. This calibration is based on a huge proprietary database. We only provide a very broad overview of the KMV model at this point, following the exposition in [103], pp. 336-338. A more detailed description of the KMV model can be found in [31] and in [32].

Within the KMV model one computes the so-called *Expected Default Frequency* (EDF) based on the firm's capital structure, the volatility of the asset returns and the current asset value in three stages.

First, KMV uses an iterative procedure to estimate the asset value and the volatility of asset returns. Their method is based on the Merton approach of modeling equity as a Call option on the underlying assets of the firm with the firm's liabilities as the strike price. Using this property of equity, one can derive the underlying asset value and asset volatility from the implied market value, the volatility of equity and the book value of liabilities.

Recall that in the classical Merton model the default probability of a given firm is determined by the probability that the asset value V_1 in one year lies below the threshold value B representing the firm's debt. Hence the default probability PD_{Merton} in the Merton model is a function of the current asset value V_0, the asset value's annualized mean μ_V and volatility σ_V, and the threshold B. With lognormally distributed asset values, this leads to a default probability (for a one year time horizon) of the form[5]

$$\text{PD}_{Merton} = 1 - \Phi\left(\frac{\ln(V_0/B) + (\mu_V - \frac{1}{2}\sigma_V^2)}{\sigma_V}\right). \tag{3.11}$$

Since asset values are not necessary lognormal, the above relationship between asset value and default probability may not be an accurate description of empirically observed default probabilities.

---

[5] Compare equation (3.4).

30   3 The Merton Model

The EDF represents an estimated probability that a given firm will default within one year. In the KMV model, the EDF is slightly different but has a similar structure as the default probability of the Merton model. The function $1 - \Phi$ in the above formula is replaced by some decreasing function which is estimated empirically in the KMV model. Moreover, the threshold value B is replaced by a new default threshold $\tilde{B}$ representing the structure of the firm's liabilities more closely, and the argument of the normal distribution function in (3.11) is replaced by a slightly simpler expression.[6] Therefore, KMV computes, in a second step, the *distance-to-default* (DD) as

$$\mathrm{DD} := (V_0 - \tilde{B})/(\sigma_V V_0). \tag{3.12}$$

It represents an approximation of the argument of (3.11), since μ_V and σ_V^2 are typically small and since $\ln V_0 - \ln \tilde{B} \approx (V_0 - \tilde{B})/V_0$.

In the last step, the DD is mapped to historical default events to estimate the EDF. In the KMV model, it is assumed that firms are homogeneous in default probabilities for equal DDs. The mapping between DD and EDF is determined empirically based on a database of historical default events. The estimated average EDF is then used as a proxy for the probability of default.

3.3.2 The CreditMetrics Model

The CreditMetrics model developed by JPMorgan and the RiskMetrics Group (see [111]) also descends from the Merton model. It deviates from the KMV model mainly through the determination of the default probability of a given firm by means of rating classes. Changes in portfolio value are only related to credit migrations of the single obligors, including both up- and downgrades as well as defaults. While in the KMV model, borrower specific default probabilities are computed, CreditMetrics assumes default and migration probabilities to be constant within the rating classes. Each firm is assigned to a certain credit-rating category at a given time period. The number of rating classes is finite and rating classes are ordered by credit quality including also the default class. One then determines the credit migration probabilities, i.e. the probabilities of moving from one rating class to another in a given time (typically one year). These probabilities are typically presented in form of a rating-transition probability matrix. The current credit rating is then assumed to determine the default probability completely which can thus be derived from the rating-transition probability matrix. Having assigned every borrower to a certain rating class and having determined the rating-transition matrix as well as the expectation and volatility of the recovery rate, the distribution of the portfolio loss can be simulated. When embedding the CreditMetrics model in an asset-value model of the Merton type, this can be achieved as sketched in Remark 3.1.1.

---
[6] Compare [103], pages 336/337.

4
The Asymptotic Single Risk Factor Model

As already mentioned in Section 2.1, the Basel Committee on Banking Supervision released a *Revised Framework*, also called *Basel II*, in June 2004. The supervisory rules are intended to provide guidelines for the supervisory authorities of the individual nations such that they can implement them in a suitable way for their banking system. Basel II provides details for adopting more risk sensitive minimal capital requirements, so-called *regulatory capital* (Pillar 1). Moreover it lays out principles for the *supervisory review process* of capital adequacy (Pillar 2) and seeks to establish *market discipline* by enhancing transparency in banks' financial reporting (Pillar 3). The Revised Framework incorporates new developments in credit risk management as it is more flexible and risk sensitive than the former Basel I accord. Moreover, within Basel II banks may opt for the *standard approach,* which is quite conservative with respect to capital charge, and the more advanced, so-called *Internal Ratings Based* (IRB) approach when calculating regulatory capital for credit risk. Financial institutions that opt for the IRB approach are allowed to use their own internal credit risk measures as inputs to the capital calculation whenever these are approved by the supervisory authorities. Therefore banks have to prove that certain conditions concerning the method and transparency are fulfilled. In the IRB approach banks are allowed to determine the borrower's default probabilities using their own methods while those using the advanced IRB approach are further permitted to provide own estimates of LGD and EAD. The Basel II risk weight formulas then translate these risk measures into risk weights and regulatory capital requirements which are intended to ensure that unexpected losses can be covered up to a certain confidence level, prescribed by the supervisors.

The risk weight formulas for the computation of regulatory capital for unexpected losses are based on the so-called *Asymptotic Single Risk Factor* (ASRF) model developed by the Basel Committee and in particular by [72]. This model was constructed in a way to ensure that the capital required for any risky loan should not depend on the particular portfolio decomposition it is added to. This so-called *portfolio invariance* was necessary for reasons of applicability relying on straightforward and fast computations of capital requirements. However, portfolio invariance comes along with some drawbacks as it makes recognition of diversification effects very difficult. Judging whether a loan fits well into an existing portfolio requires the knowledge of the portfolio decomposition and therefore contradicts portfolio invariance. Thus the ASRF model is based on the assumption of a well diversified portfolio. We will see later on that actually no real bank can exactly fulfill this assumption. Therefore banks are expected to account for this existence of *concentration risks* in Pillar 2. Models for this task will be presented in Part II.

In the following sections we will first present the general framework of the ASRF model and then discuss its shortcomings and some practical aspects.

4.1 The ASRF Model

The ASRF model developed by [72] is based on the law of large numbers. An ordinary portfolio consists of a large number of exposures of different sizes. When exposure sizes are equally distributed, the idiosyncratic risk associated with every single exposure is almost *diversified away;* meaning that the idiosyncratic risks cancel out one-another. Note that this requires a very huge portfolio. If the portfolio consists of only two exposures of the same size then idiosyncratic risk is quite significant. When idiosyncratic risk is diversified away, only systematic risk affecting many exposures remains. Such a portfolio is also called *infinitely fine-grained*. It is needless to say that such perfectly fine-grained portfolios do not exist in practice. Real bank portfolios have a finite number of obligors and lumpy distributions of exposure sizes. The asymptotic assumption might be approximately valid for some of the largest bank portfolios, but clearly would be much less satisfactory for portfolios of smaller or more specialized institutions. Thus any capital charges computed under the assumption of an asymptotically fine-grained portfolio must underestimate the required capital for a real finite portfolio. Therefore, banks have to account for this non-diversified idiosyncratic risk under Pillar 2. We will address this effect of *name concentration* or *granularity* later on in Chapter 9. In the ASRF model all systematic risks in a portfolio are modeled by a single systematic risk factor representing for example the overall economic climate. Now we already have stated the two main assumptions of the ASRF model which are also eponymous for the model.

Assumption 4.1.1

1. *Portfolios are infinitely fine-grained, i.e. no exposure accounts for more than an arbitrarily small share of total portfolio exposure.*
2. *Dependence across exposures is driven by a single systematic risk factor.*

4.1 The ASRF Model

Under these assumptions it is possible to compute the sum of the expected and unexpected losses associated with each loan. This is done by computing the conditional expected loss for each exposure, i.e. the expected loss of the exposure given an appropriately conservative value of the single systematic risk factor.

Consider the portfolio of N risky loans described in Section 1.2. We denote the systematic risk factor by X. For now, X can also be multivariate. Then the risk factors can, for example, represent some industrial or geographical regions. It is assumed that all dependence across obligors is due to the sensitivity to this common set of risk factors. The unconditional default probability PD_n can be derived from the conditional default probability $\text{PD}_n(X)$ by integrating over all possible realizations of the systematic risk factor X and thereby averaging out the systematic risk. Conditional on X the default probabilities and also the portfolio's credit risk is purely idiosyncratic to the individual exposures. Note that this general framework is in line with most of the widely applied industry models as for examples the ones presented in Chapter 3. [70] presents a detailed derivation of these models in the above framework.

Denote by EAD_n the exposure at default to obligor n which is assumed to be known and non-stochastic and let LGD_n be obligor n's percentage loss in the event of default which we permit to be negative to incorporate also short positions. As before let D_n denote the default indicator variable of obligor n.

Assumption 4.1.2
Assume that the variables $U_n \equiv \text{LGD}_n \cdot D_n$, for $n = 1, \ldots, N$, are bounded in the interval $[-1, 1]$ and are mutually independent conditional on the systematic factor X.[1]

We express loss not in absolute value but in percentage of total exposure. Therefore, denote the exposure share of obligor n by

$$s_n = \frac{\text{EAD}_n}{\sum_{n=1}^{N} \text{EAD}_n}.$$

Then the *portfolio loss ratio* L is given by

$$L = \sum_{n=1}^{N} D_n \cdot \text{LGD}_n \cdot s_n.$$

The first condition in Assumption of 4.1.1 is satisfied when the sequence of positive constant exposures EAD_n satisfies the following conditions.[2]

Assumption 4.1.3
(a) $\sum_{n=1}^{N} \text{EAD}_n \uparrow \infty$ *and*

---
[1] Compare [72], Assumption ($\mathcal{A}$-1), p. 204.
[2] Compare [72], Assumption ($\mathcal{A}$-2), p. 205.

(b) *there exists a positive ξ such that the largest exposure share is of order $\mathcal{O}(N^{-(1/2+\xi)})$. Hence it shrinks to zero as the number of exposures in the portfolio increases to ∞.*

Under these conditions the following theorem follows directly from the strong law of large numbers (see [72], Proposition 1, for a formal proof).

Theorem 4.1.4 *Under Assumptions 4.1.2 and 4.1.3 the portfolio loss ratio conditional on a realization x of the systematic risk factor X satisfies*

$$L_N - \mathbb{E}[L_N|x] \to 0 \quad \text{almost surely as } N \longrightarrow \infty.$$

(We included the subscript N in the portfolio loss $L \equiv L_N$ here only to denote the dependence on the portfolio size N.)

This is the main result on which the ASRF model is based. It mainly says that the larger the portfolio is, the more idiosyncratic risk is diversified away. In the limit the portfolio is driven solely by systematic risk. This limiting portfolio is also called *infinitely fine-grained* or *asymptotic portfolio*.

So far we have not made use of the single systematic risk factor assumption in 4.1.1. Let us first impose another assumption.[3]

Assumption 4.1.5
There is an open interval B containing the qth percentile $\alpha_q(X)$ of the systematic risk factor X and there is a real number $N_0 < \infty$ such that

(i) *for all n, $\mathbb{E}[U_n|x]$ is continuous in x on B,*
(ii) *$\mathbb{E}[L_N|x]$ is nondecreasing in x on B for all $N > N_0$, and*
(iii) *$\inf_{x \in B} \mathbb{E}[L_N|x] \geq \sup_{x \leq \inf B} \mathbb{E}[L_N|x]$ and
$\sup_{x \in B} \mathbb{E}[L_N|x] \leq \inf_{x \geq \sup B} \mathbb{E}[L_N|x]$ for all $N > N_0$.*

These assumptions guarantee that the neighborhood of the qth quantile of $\mathbb{E}[L_N|X]$ is associated with the neighborhood of the qth quantile of the systematic factor X. Otherwise the tail quantiles of the loss distribution would depend in a complex way on the behavior of the conditional expected loss for each borrower. This problem could also be avoided by requiring $\mathbb{E}[U_n|x]$ to be nondecreasing in x for all n. This, however, would exclude the possibility that some U_n are negatively associated with the factor X. Such a counter-cyclical behavior occurs for example in the case of hedging instruments.

Under Assumptions 4.1.1 (2) and 4.1.5 it can be shown that the following Theorem holds.

Theorem 4.1.6 *Suppose Assumptions 4.1.1 (2) and 4.1.5 hold. Then for $N > N_0$ we have*

$$\alpha_q(\mathbb{E}[L_N|X]) = \mathbb{E}[L_N|\alpha_q(X)], \tag{4.1}$$

where $\alpha_q(\mathbb{E}[L_N|X])$ denotes the qth-quantile of the random variable $\mathbb{E}[L_N|X]$.

---
[3] Compare [72], Assumption (*A*-4), p. 207.

For a proof see [72], Proposition 4. This relation is the core of the Basel risk weight formulas. It presents a portfolio invariant rule to determine capital requirements by taking the exposure-weighted average of the individual assets' conditional expected losses.

4.2 The IRB Risk Weight Functions

If we now want to translate the theoretical ASRF model, presented in Section 4.1, into a practical model for the calculation of regulatory capital, we first have to find a way to derive conditional default probabilities. This is done by an adaption of Merton's model. Recall that in the Merton model an obligor defaults if its asset value falls below a threshold given by its obligations. Within the Merton model the asset value is described by a normally distributed random variable. The Basel Committee adopted the assumption of normally distributed risk factors.

Hence, the ASRF model can be described as a factor model such that the return on the firm's assets is of the form (3.6), i.e.

$$r_n = \sqrt{\varrho_n} \cdot X + \varepsilon_n, \quad n = 1, \ldots, N,$$

with normally distributed systematic risk factor X and idiosyncratic shocks ε_n. Here r_n denotes the log-asset return of obligor n and ϱ_n captures the correlation between r_n and the single risk factor X. The conditional default probabilities are thus given by equation (3.8), i.e. by

$$\mathrm{PD}_n(X) = \Phi\left(\frac{\Phi^{-1}(\mathrm{PD}_n) - \sqrt{\varrho_n} \cdot X}{\sqrt{1 - \varrho_n}}\right). \qquad (4.2)$$

Choosing a realization x for the systematic risk factor equal to the qth quantile $\alpha_q(X)$ of the systematic risk factor and taking into account, that X is assumed to be normally distributed, we obtain

$$\mathrm{PD}_n(\alpha_q(X)) = \Phi\left(\frac{\Phi^{-1}(\mathrm{PD}_n) + \sqrt{\varrho_n} \cdot \Phi^{-1}(q)}{\sqrt{1 - \varrho_n}}\right). \qquad (4.3)$$

The variable ϱ_n describes the degree of the obligor's exposure to the systematic risk factor by means of the asset correlation and is determined by the borrower's asset class. In the Revised Framework the asset value correlation for a given asset class, represented by the according unconditional default probability PD_n, is given by

$$\varrho_n = 0.12 \cdot \frac{1 - \exp(-50 \cdot \mathrm{PD}_n)}{1 - \exp(-50)} + 0.24 \cdot \left(1 - \frac{1 - \exp(-50 \cdot \mathrm{PD}_n)}{1 - \exp(-50)}\right). \qquad (4.4)$$

The behavior of the correlation coefficient ϱ_n in dependence of the unconditional default probabilities is shown in Figure 4.1. The dashed lines denote the upper and lower bound of the correlation coefficient which are given by 24% and 12%, respectively. Correlations between these limits are given by the exponential weighting function (4.4) describing the dependence on the PD.

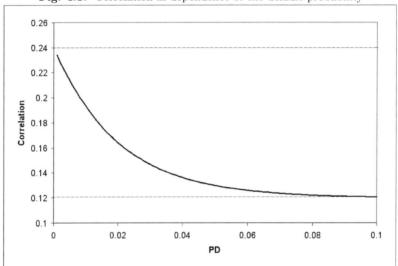

Fig. 4.1. Correlation in dependence of the default probability

Having assigned an unconditional default probability PD_n to each obligor and having computed the conditional default probabilities $PD_n(X)$ via formula (4.3), we can return to the computation of regulatory capital in the ASRF model. The ASRF model suggests to determine capital requirements by taking the exposure-weighted average of the individual assets' conditional expected losses. The EL of loan n is given by

$$EL_n = PD_n \cdot EAD_n \cdot LGD_n.$$

The LGD used to calculate an exposure's conditional expected loss must also reflect financial distress. Therefore, Basel II uses a so-called *downturn* LGD. The intuition behind this concept is that, during times of financial distress or economic downturn, losses on defaulted loans are higher than under normal conditions because, for example, collateral values may decline. One method to obtain downturn LGDs, would be to construct a mapping function similar to the derivation of the conditional default probability, to transform average LGDs into downturn LGDs. The Basel Committee, however, decided that a single supervisory LGD mapping function would be inappropriate due to the evolving nature of bank practices in the area of LGD estimations. Therefore, banks opting for the advanced IRB approach are allowed to use their own internal methods to derive downturn LGDs.

The expected loss conditional on the qth quantile of the systematic risk factor X can be estimated by

$$\mathbb{E}[L_n|\alpha_q(X)] = \text{PD}_n(\alpha_q(X)) \cdot \text{LGD}_n \cdot \text{EAD}_n$$

where LGD_n denotes the downturn LGD of obligor n. Since under the Revised Framework capital requirements are computed on UL-only basis, the expected loss has to be subtracted to derive the Basel II capital requirements. Since capital requirements in the Revised Framework are expressed as a percentage of total exposure, the risk weighted assets need to be multiplied by EAD_n and the reciprocal of the minimum capital ratio of 8%, i.e. by the factor 12.5, since the minimum capital requirements for obligor n are later on computed as 8% of the risk weighted assets of obligor n. Thus the *risk weighted assets* RWA_n of obligor n are given by

$$\text{RWA}_n = 12.5 \cdot \text{EAD}_n \cdot \text{MA}_n$$

$$\cdot \left(\text{LGD}_n \cdot \Phi \left(\frac{\Phi^{-1}(\text{PD}_n) + \sqrt{\varrho_n} \cdot \Phi^{-1}(q)}{\sqrt{1-\varrho_n}} \right) - \text{LGD}_n \cdot \text{PD}_n \right),$$

where LGD_n denotes the downturn LGD of obligor n. Here MA_n is an obligor specific *maturity adjustment* term given by

$$\text{MA}_n = \frac{1 + (M_n - 2.5) \cdot b(\text{PD}_n)}{1 - 1.5 \cdot b(\text{PD}_n)} \tag{4.5}$$

with effective maturity M_n of loan n and where

$$b(\text{PD}_n) = (0.11852 - 0.05478 \cdot \ln(\text{PD}_n))^2$$

is a smoothed regression maturity. Such a maturity adjustment has to be applied since credit portfolios consist of instruments of very different maturities. It is well known that long-term loans are much riskier than short-term credits. This should also be reflected in the risk weight functions. The maturity adjustment (4.5) depends on the default probabilities since loans with high PDs usually have a lower market value today than loans with low PDs with the same face value. The concrete form of the Basel maturity adjustment stems from a mark-to-market credit risk model, similar to the KMV model that is consistent with the ASRF model, and which has been smoothed by a statistical regression model. This regression function has been chosen in a way to ensure that the maturity adjustments are linear and increasing in M. As a function of M, the maturity adjustment is decreasing in the default probabilities. For a one year maturity the MA yields the value 1.

38   4 The Asymptotic Single Risk Factor Model

The *minimal capital requirements* MCR_n for obligor n are then given by 8% of RWA_n. Hence the factor 12.5 in the formula for the RWA cancels out with the 8%. For the total portfolio we thus obtain minimal capital requirements of

$$\text{MCR} = 0.08 \cdot \sum_{n=1}^{N} \text{RWA}_n .$$

Figure 4.2 shows the dependence of the minimum capital requirements MCR on the unconditional default probability PD_n for a homogeneous portfolio where all exposure shares equal $1/N$, default probabilities are constant throughout the portfolio and LGDs are constant equal to 45%. For a $\text{PD} \geq 2\%$ the MCRs increase proportional to the PDs.

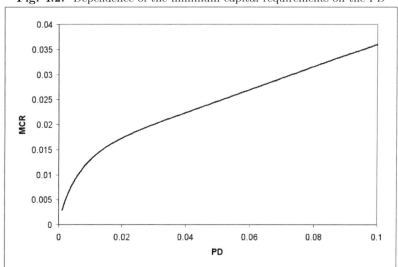

Fig. 4.2. Dependence of the minimum capital requirements on the PD

4.3 The Loss Distribution of an Infinitely Granular Portfolio

Now we want to compute the loss distribution function in an infinitely granular portfolio. Therefore we denote as before the exposure at default of obligor n by EAD_n and the exposure share of obligor n by

$$s_n = \text{EAD}_n / \sum_{k=1}^{N} \text{EAD}_k .$$

4.3 The Loss Distribution of an Infinitely Granular Portfolio

The loss given default of obligor n is denoted by LGD_n. Then, for a portfolio of N loans, the percentage portfolio loss L_N is given by

$$L_N = \sum_{n=1}^{N} s_n \cdot \text{LGD}_n \cdot D_n$$

where D_n is the default indicator variable of obligor n which is Bernoulli $\text{Bern}(1; \text{PD}_n(X))$ distributed. Here, as before, X denotes the systematic factor which is $\mathcal{N}(0,1)$ distributed and $\text{PD}_n(X)$ is the conditional default probability of obligor n which is given by equation (4.2). Under the assumptions of the ASRF model, Theorem 4.1.4 states that

$$\mathbb{P}\left(\lim_{N \to \infty} (L_N - \mathbb{E}[L_N|X]) = 0\right) = 1.$$

For simplicity we now assume a homogeneous portfolio in the sense that all obligors have the same default probability $\text{PD}_n = \text{PD}$, for $n = 1, \ldots, N$, and we assume that $\text{LGD}_n = 100\%$ for $n = 1, \ldots, N$. Then the conditional default probability can be computed as

$$\text{PD}(X) = \Phi\left(\frac{\Phi^{-1}(p) - \sqrt{\varrho} \cdot X}{\sqrt{1-\varrho}}\right), \tag{4.6}$$

where ϱ is the correlation between obligor n and the risk factor X. Note that, since default probabilities are constant, the correlations are also constant. Then we obtain

$$\mathbb{E}[L_N|X] = \sum_{n=1}^{N} s_n \cdot \mathbb{E}[D_n|X] = \text{PD}(X),$$

since the sum over the exposure shares equals 1. Hence, in the limit as $N \to \infty$, the percentage portfolio loss L_N tends to the conditional default probability

$$L_N \to \text{PD}(X) \quad \text{almost surely as } N \to \infty. \tag{4.7}$$

Thus, in an infinitely fine-grained portfolio, the conditional default probability $\text{PD}(X)$ describes the fraction of defaulted obligors. Denote the percentage number of defaults in such a portfolio by L. Then we can compute the probability density function of L as follows. For every $0 \leq x \leq 1$ we have

$$F_L(x) \equiv \mathbb{P}[L \leq x] = \mathbb{P}[\text{PD}(X) \leq x]$$

$$= \mathbb{P}\left[-X \leq \frac{1}{\sqrt{\varrho}}\left(\sqrt{1-\varrho} \cdot \Phi^{-1}(x) - \Phi^{-1}(\text{PD})\right)\right]$$

$$= \Phi\left(\frac{1}{\sqrt{\varrho}}\left(\sqrt{1-\varrho} \cdot \Phi^{-1}(x) - \Phi^{-1}(\text{PD})\right)\right).$$

Differentiating with respect to x gives the probability density function

Fig. 4.3. Densities of the distribution of the percentage number of defaults for different values of ϱ and PD

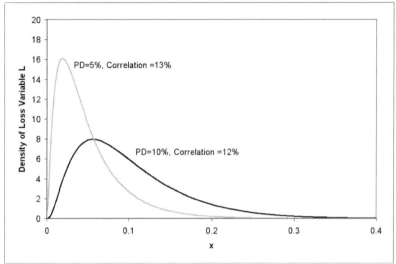

$$f_L(x) = \frac{\partial F_L(x)}{\partial x} =$$

$$= \sqrt{\frac{1-\varrho}{\varrho}} \cdot \frac{1}{\sqrt{2\pi}} \exp\left(-\frac{1}{2\varrho}\left(\sqrt{1-\varrho}\cdot\Phi^{-1}(x) - \Phi^{-1}(\text{PD})\right)^2\right) \cdot \frac{\partial}{\partial x}\left[\Phi^{-1}(x)\right]$$

$$= \sqrt{\frac{1-\varrho}{\varrho}} \exp\left(-\frac{1}{2\varrho}\left(\sqrt{1-\varrho}\cdot\Phi^{-1}(x) - \Phi^{-1}(\text{PD})\right)^2\right) \cdot \exp\left(\frac{1}{2}\left(\Phi^{-1}(x)\right)^2\right).$$

In the correlation-free case when $\varrho = 0$, the portfolio loss follows a binomial distribution and the density f_L degenerates to a distribution concentrated in the point PD, the *Dirac measure* at PD. In the perfect correlation case when $\varrho = 1$, the percentage portfolio loss no longer depends on the portfolio size N and is Bern(1, PD) distributed. The density f_L degenerates to a distribution concentrated in the points PD and $1-$PD. Two other extreme scenarios do occur when the default probability is PD $= 0$, meaning that all obligors survive almost surely, or when PD $= 1$, in which case all obligors default almost surely.

For realistic values of PD and ϱ, Figure 4.3 shows the graph of the density function f_L. The grey graph corresponds to a default probability of 5% and a correlation of $\varrho = 13\%$. The black graph belongs to values PD $= 10\%$ and $\varrho = 12\%$. Here the correlation values have been computed via the IRB correlation function (4.4).

4.3 The Loss Distribution of an Infinitely Granular Portfolio

Using Theorem 4.1.6 we can easily compute the quantiles of the percentage portfolio loss. Figure 4.4 shows the 95% quantile of the percentage portfolio loss of an infinitely granular portfolio with PD = 10% and $\varrho = 12\%$. Accordingly, Figure 4.5 shows the 95% quantile of the percentage portfolio loss of an infinitely granular portfolio with PD = 5% and $\varrho = 13\%$. We chose the 95% quantile only for illustrative purposes. Of course, in practice the 99.9% or even the 99.98% quantiles are used more frequently. These values, however, would be quite hard to recognize in the figures since they are far in the tail of the distributions.

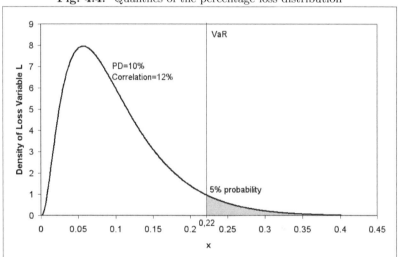

Fig. 4.4. Quantiles of the percentage loss distribution

We can now easily compute the economic capital EC_q at level q for an infinitely fine-grained portfolio with respect to PD and ϱ. For the above portfolio with $\varrho = 13\%$ and PD = 5% (now denoted Portfolio 1) we obtain a 99.5% economic capital value of $EC_{99.5} = 17.1\%$ and a 99.9% economic capital value of $EC_{99.9} = 23.4\%$. For the values $\varrho = 12\%$ and PD = 10% (Portfolio 2) we derive accordingly $EC_{99.5} = 24.0\%$ and $EC_{99.9} = 31.2\%$. The unexpected loss of Portfolio 1 equals 4.0% and for Portfolio 2 the unexpected loss is 6.4%. Hence we see that the unexpected loss risk capital is usually much lower than the economic capital. This also underlines the fact, we already stated in Section 2.5, that holding only the unexpected loss reserve as a risk capital might not be appropriate in situations of financial distress.

Fig. 4.5. Quantiles of the percentage loss distribution

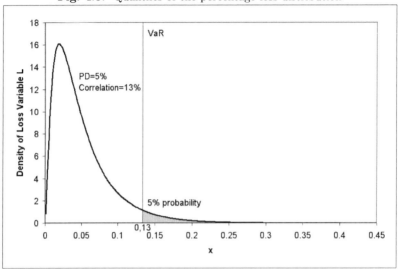

5

Mixture Models

Current credit risk models can be divided into two classes. On the one hand there are *latent variable* or *threshold models* where default occurs when a latent variable (e.g. the firm's asset value) falls below some threshold. Examples for this class of credit risk models are the theoretical Merton model and, based on this, also important industry models like KMV's Portfolio Manager or the CreditMetrics Model. We studied these types of models in some detail in Chapter 3. On the other hand there are *mixture models* where the default probabilities of different obligors are assumed to depend on some common economic factors. These models can be treated as two stage models. Conditional on the realization of economic factors the individual default probabilities are assumed to be independent whereas they can be unconditionally dependent. The conditional default probabilities are modeled as random variables with some mixing distribution which is specified in a second step. This class of credit risk models will be treated in this chapter. A prominent example of such a mixture model is the CreditRisk$^+$ model which we will discuss in more detail in Chapter 6.

5.1 Bernoulli and Poisson Mixture Models

In this section we present the *Bernoulli mixture model* and the *Poisson mixture model* as the most prominent examples of mixture models. In the case of the Bernoulli mixture model the default probabilities are Bernoulli distributed conditional on the set of economic factors where the mixing variable, the Bernoulli parameter, is random. Formally, the following definition holds.

Definition 5.1.1 (Bernoulli Mixture Model)
Given some $K < N$ and a K-dimensional random vector $X = (X_1, \ldots, X_K)$, the random vector $D = (D_1, \ldots, D_N)$ of default indicators follows a *Bernoulli mixture model* with factor vector X, if there are functions $p_n : \mathbb{R}^K \to [0, 1]$,

$1 \leq n \leq N$, such that conditional on a realization x of X the components of D are independent Bernoulli random variables with

$$\mathbb{P}(D_n = 1 | X = x) = p_n(x).$$

For some $d = (d_1, \ldots, d_N) \in \{0,1\}^N$, we then obtain for the joint default probability conditional on a realization x of X

$$\mathbb{P}(D_1 = d_1, \ldots, D_N = d_N | X = x) = \prod_{n=1}^{N} p_n(x)^{d_n} \cdot (1 - p_n(x))^{1-d_n}.$$

Integration over the distribution of the factor vector X, the *mixing distribution*, then yields the unconditional distribution of the default indicator D. In particular, we obtain for the unconditional default probability PD_n of an individual obligor n,

$$\text{PD}_n = \mathbb{P}(D_n = 1) = \mathbb{E}[p_n(X)].$$

Let

$$M = \sum_{n=1}^{N} D_n$$

denote the number of defaulted companies up to time T.

Since default is a rare event and thus PD_n is very small, we can approximate the Bernoulli distributed default indicator variable by a Poisson random variable. We will use this *Poisson approximation* later on again when we studied the CreditRisk$^+$ model in Chapter 6. In the context of mixture models this leads to the notion of Poisson mixture models.

Definition 5.1.2 (Poisson Mixture Model)
Given some K and X, as in Definition 5.1.1. The random vector of default indicators $\tilde{D} = (\tilde{D}_1, \ldots, \tilde{D}_N)$ follows a *Poisson mixture model* with factor vector X, if there are functions $\lambda_n : \mathbb{R}^K \to (0, \infty)$, $1 \leq n \leq N$, such that conditional on a realization x of the factor vector X, the random vector $\tilde{D}$ is a vector of independent Poisson random variables with intensities $\lambda_n(x)$.

Note that in this setting the default indicator variable $\tilde{D}_n$ of obligor n can take values in $\{0, 1, 2, \ldots\}$, indicating that an obligor can default more than once in the given time period. Hence $\tilde{D}_n$ counts the number of defaults of obligor n up to time T. For small default intensities λ_n the random variable

$$\tilde{M} = \sum_{n=1}^{N} \tilde{D}_n$$

corresponds approximately to the number of defaulted companies. Conditional on a realization x of the factor vector, $\tilde{M}$ is just the sum of independent Poisson variables and thus its distribution satisfies

5.1 Bernoulli and Poisson Mixture Models

$$\mathbb{P}(\tilde{M} = k | X = x) = \exp\left(-\sum_{n=1}^{N} \lambda_n(x)\right) \cdot \frac{\left(\sum_{n=1}^{N} \lambda_n(x)\right)^k}{k!}.$$

It is possible to relate the functions p_n and λ_n by setting the random variable $D_n = \mathbf{1}_{\{\tilde{D}_n \geq 1\}}$. We then obtain

$$\begin{aligned} p_n(X) &= \mathbb{P}(D_n = 1 | X) \\ &= \mathbb{P}(\tilde{D}_n \geq 1 | X) \\ &= 1 - \mathbb{P}(\tilde{D}_n = 0 | X) = 1 - \exp(-\lambda_n(X)). \end{aligned}$$

For reasons of simplification we consider in the following only exchangeable models.[1] In this context we obtain identical functions $p_n(X) = p(X)$ and $\lambda_n(X) = \lambda(X)$ for all obligors. The conditional probabilities of the number of defaults in the Bernoulli mixture model and in the Poisson mixture model are given by

$$\mathbb{P}(\tilde{M} = k | X) = e^{-N \cdot \lambda(X)} \cdot \frac{(N \cdot \lambda(X))^k}{k!}$$

and

$$\mathbb{P}(M = k) = \binom{N}{k} \cdot p(X)^k \cdot (1 - p(X))^{N-k},$$

respectively. $p(X)$ and $\lambda(X)$ are, of course, random variables and we denote their probability distributions by G_p and G_λ, respectively. These distributions are sometimes also referred to as *the mixing distributions* of $p(X)$ and $\lambda(X)$. We obtain the unconditional probabilities of the number of defaults by integrating over the mixing distributions, i.e.

$$\mathbb{P}(\tilde{M} = k) = \int_0^\infty e^{-N \cdot y} \cdot \frac{(N \cdot y)^k}{k!} dG_\lambda(y) \tag{5.1}$$

$$\mathbb{P}(M = k) = \binom{N}{k} \int_0^1 y^k \cdot (1 - y)^{N-k} dG_p(y). \tag{5.2}$$

Note that in case of an exchangeable Bernoulli mixture model we have

$$\begin{aligned} \mathbb{P}(D_1 = d_1, \ldots, D_N = d_N | X = x) &= \prod_{n=1}^{N} p(x)^{d_n} \cdot (1 - p(x))^{1-d_n} \\ &= p(x)^{\delta_N} \cdot (1 - p(x))^{N - \delta_N}. \end{aligned}$$

with $\delta_N = \sum_{n=1}^{N} d_n$. Integrating out the condition on the factor vector X, we obtain

$$\mathbb{P}(D_1 = d_1, \ldots, D_N = d_N) = \int_0^1 y^{\delta_N} \cdot (1 - y)^{N - \delta_N} dG_p(y).$$

---

[1] See Section 1.3 for definitions.

5 Mixture Models

This is exactly what a theorem by de Finetti [39] states, namely that all exchangeable binary sequences of random variables are mixtures of Bernoulli random variables.

Theorem 5.1.3 (de Finetti's Theorem) *A binary sequence $D_1, D_2, \ldots$ is exchangeable if and only if there exists a distribution function F on $[0,1]$ such that for all $N \geq 1$ the following holds*

$$\mathbb{P}(D_1 = d_1, \ldots, D_N = d_N) = \int_0^1 y^{\delta_N} \cdot (1-y)^{N-\delta_N} \, dF(y),$$

where $\delta_N := \sum_{n=1}^N d_n$. Moreover, F is the distribution function of the limiting frequency

$$Y := \lim_{N \to \infty} \sum_{n=1}^N D_n / N, \quad \mathbb{P}(Y \leq y) = F(y),$$

while the Bernoulli distribution is obtained by conditioning on $Y = y$, i.e.

$$\mathbb{P}(D_1 = d_1, \ldots, D_N = d_N | Y = y) = y^{\delta_N} \cdot (1-y)^{N-\delta_N}.$$

Particularly for Bernoulli mixture models one can show, using the law of iterated expectations, that unconditional default probabilities of first and higher order correspond to the moments of the mixing distribution

$$\pi_k = \mathbb{P}(D_1 = 1, \ldots, D_k = 1) = \mathbb{E}[\mathbb{E}[D_1 \cdots D_k | p(X)]] = \mathbb{E}[p(X)^k].$$

The precise probability obviously depends on the assumed mixing distribution. In the following we calculate this probability explicitly for two frequently used examples of mixture models.

Example 5.1.4 (Beta-Binomial Distribution) Different choices of the distribution of $p(X)$ lead to different distributions for the number of defaults. One example is to assume a Beta distribution for $p(X)$ with density

$$g_p(y) = \frac{1}{\beta(a,b)} \cdot y^{a-1} \cdot (1-y)^{b-1}$$

and $a, b > 0$. The Beta function is given by

$$\beta(a,b) = \int_0^1 x^{a-1} \cdot (1-x)^{b-1} dx.$$

In this case, the k-th order of the unconditional default probability can be calculated as

$$\pi_k = \mathbb{E}[p(X)^k] = \int_0^1 y^k dG_p(y)$$

$$= \frac{1}{\beta(a,b)} \int_0^1 y^k \cdot y^{a-1} \cdot (1-y)^{b-1} dy$$

$$= \frac{1}{\beta(a,b)} \cdot \beta(k+a,b)$$

$$= \prod_{j=0}^{k-1} \frac{a+j}{a+b+j}$$

where we used the recursion formula

$$\frac{\beta(a+1,b)}{\beta(a,b)} = \frac{a}{a+b}.$$

Similarly, we obtain from equation (5.2), that the number of defaults follows a so-called Beta-Binomial distribution, i.e.

$$\mathbb{P}(M=k) = \binom{N}{k} \frac{\beta(a+k,b+N-k)}{\beta(a,b)}.$$

Example 5.1.5 (Negative Binomial Distribution) A natural choice for the mixing distribution for $\lambda(X)$ is the Gamma distribution with density

$$g_\lambda(y) = b^a \cdot y^{a-1} \cdot e^{-by}/\Gamma(a)$$

where $a, b > 0$ are constants. Recall that the Gamma function is given by

$$\Gamma(\alpha) = \int_0^\infty x^{a-1} \cdot e^x dx.$$

The distribution of the number of defaults can then be obtained from equation (5.1) as

$$\mathbb{P}(\tilde{M}=k) = \int_0^\infty e^{-Ny} \cdot \frac{N^k y^k}{k!} \cdot b^a \cdot y^{a-1} \cdot e^{-by}/\Gamma(a) dy$$

$$= \frac{N^k b^a}{k! \Gamma(a)} \int_0^\infty y^{k+a-1} \cdot e^{-(b+N)y} dy.$$

Substitution of $u = (b+N) \cdot y$ yields

$$\mathbb{P}(\tilde{M}=k) = \frac{N^k b^a}{k! \Gamma(a)} \int_0^\infty (b+N)^{-(k+a)+1} \cdot u^{k+a-1} \cdot e^{-u} du$$

$$= \frac{N^k b^a}{k! \Gamma(a)} \cdot (b+N)^{-(k+a)+1} \cdot \Gamma(a+k)$$

$$= \left(1 - \frac{b}{b+N}\right)^k \cdot \left(\frac{b}{b+N}\right)^a \cdot \frac{\Gamma(a+k)}{\Gamma(a)\Gamma(k+1)}$$

where we used $\Gamma(k+1) = k! \cdot \Gamma(1)$. This corresponds to the Negative Binomial distribution $NB\left(a, \frac{b}{N+b}\right)$.

5.2 The Influence of the Mixing Distribution on the Loss Distribution

For credit risk management it is important to know the influence of a particular choice of the mixing distribution on the tails of the loss distribution. In the mixture models framework, [59] explore the role of the mixing distribution of the factors and find that the tail of the mixing distribution essentially determines the tail of the overall credit loss distribution. In particular, it can be shown that in large portfolios the quantile of the loss distribution is determined by the quantile of the mixing distribution which is the main result of Theorem 5.2.1 below. First remember that the credit portfolio loss is given by

$$L_N = \sum_{n=1}^{N} \text{LGD}_n \cdot D_n,$$

where we assume that for all obligors n the random loss given defaults LGD_n have common mean ELGD and variance $\text{VLGD}^2 < \infty$. The following theorem and its proof are due to [59], Proposition 4.5.

Theorem 5.2.1 *Given some factor vector X, let $\alpha_q(p(X))$ be the qth quantile of the mixing distribution G_p of the conditional default probability $p(X)$, i.e.*

$$\alpha_q(p(X)) = \inf\{x \in \mathbb{R} : G_p(x) \geq q\},$$

and define $\alpha_q(L_N)$ to be the qth quantile of the credit loss distribution in a portfolio of N obligors. Assume that the quantile function $q \mapsto \alpha_q(p(X))$ is continuous in q. Then

$$\lim_{N \to \infty} \frac{1}{N} \alpha_q(L_N) = \text{ELGD} \cdot \alpha_q(p(X)).$$

Proof. Note first that, conditional on $X = x$, the individual losses $\text{LGD}_n \cdot D_n$ form an iid sequence with mean $\text{ELGD} \cdot p(x)$. From the law of large numbers and the central limit theorem we obtain

$$\lim_{N \to \infty} \mathbb{P}\left(\frac{1}{N} \sum_{n=1}^{N} \text{LGD}_n \cdot D_n \leq c \,|\, X = x\right) = \begin{cases} 1, & \text{if } \text{ELGD} \cdot p(x) < c \\ 1/2, & \text{if } \text{ELGD} \cdot p(x) = c \\ 0, & \text{if } \text{ELGD} \cdot p(x) > c. \end{cases} \quad (5.3)$$

This yields for any $\varepsilon > 0$

5.2 The Influence of the Mixing Distribution on the Loss Distribution

$$\limsup_{m \to \infty} \mathbb{P}(L_N \leq N \cdot (\text{ELGD} \cdot \alpha_q(p(X)) - \varepsilon))$$

$$= \limsup_{N \to \infty} \int_0^1 \mathbb{P}\left(L_N \leq N \cdot (\text{ELGD} \cdot \alpha_q(p(X)) - \varepsilon)|p(X) = y\right) dG_p(y)$$

$$\leq \int_0^1 \limsup_{N \to \infty} \mathbb{P}\left(L_N \leq N \cdot (\text{ELGD} \cdot \alpha_q(p(X)) - \varepsilon)|p(X) = y\right) dG_p(y)$$

$$= \int_0^1 \left(\mathbf{1}_{\{y < \alpha_q(p(X)) - \varepsilon/\text{ELGD}\}} + \frac{1}{2} \cdot \mathbf{1}_{\{y = \alpha_q(p(X)) - \varepsilon/\text{ELGD}\}}\right) dG_p(y)$$

$$\leq \int_0^1 \mathbf{1}_{\{y \leq \alpha_q(p(X)) - \varepsilon/\text{ELGD}\}} dG_p(y)$$

$$= G_p\left(\alpha_q(p(X)) - \varepsilon/\text{ELGD}\right) < q$$

where the first inequality follows from an application of Fatou's Lemma. Moreover, we used the law of large numbers and the definition of the quantile. Analogously we have,

$$\liminf_{m \to \infty} \mathbb{P}\left(L_N \leq N(\text{ELGD} \cdot \alpha_q(p(X)) + \varepsilon)\right)$$

$$\geq \mathbb{P}(p(X) < \alpha_q(p(X)) + \varepsilon/\text{ELGD}) > q$$

which implies for N large enough that

$$N(\text{ELGD} \cdot \alpha_q(p(X))) - \varepsilon) \leq \alpha_q(L_N) \leq N(\text{ELGD} \cdot \alpha_q(p(X)) + \varepsilon)$$

and thus the assertion. □

Remark 5.2.2 Note that Theorems 4.1.4 and 4.1.6 imply the same result when applied in the framework of exchangeable mixture models. That is in the mixture models setting we have to replace the systematic risk factor X of the ASRF model by the mixing variable $p(X)$. Due to Theorem 4.1.4 we then obtain that, conditional on a realization $p(x)$ of $p(X)$,

$$\lim_{N \to \infty} (L_N - \mathbb{E}[L_N|p(x)]) = 0.$$

Together with Theorem 4.1.6 this implies that

$$\lim_{N \to \infty} \frac{1}{N} \alpha_q(L_N) = \lim_{N \to \infty} \frac{1}{N} \alpha_q(\mathbb{E}[L_N|p(X)])$$

$$= \lim_{N \to \infty} \frac{1}{N} \sum_{n=1}^N \text{ELGD} \cdot \mathbb{E}[D_n|\alpha_q(p(X))]$$

$$= \text{ELGD} \cdot \alpha_q(p(X))$$

which is just the claim of the above theorem.

5.3 Relation Between Latent Variable Models and Mixture Models

In this section we discuss the relationship between latent variable models and Bernoulli mixture models based on work by [61]. Assume the general credit portfolio setting of Section 1.2 and recall the Definition 3.0.2 of a latent variable model from Chapter 3. Although both types of models seem to be quite different, [61] provide a condition under which a latent variable model can be written as a Bernoulli mixture model.[2]

Definition 5.3.1 (Conditional Independence Structure)
A latent variable vector V has a K-dimensional *conditional independence structure* with conditioning variable X, if there is some $K < N$ and a K-dimensional random vector $X = (X_1, \ldots, X_K)$ such that conditional on X the random variables $(V_n)_{1 \leq n \leq N}$ are independent.

Proposition 5.3.2 *Consider an N-dimensional latent variable vector V and a K-dimensional ($K < N$) random vector X. Then the following are equivalent.*

(i) *V has K-dimensional conditional independence structure with conditioning variable X.*
(ii) *For any choice of thresholds b_1^n, $1 \leq n \leq N$, the default indicators $D_n = \mathbf{1}_{\{V_n \leq b_1^n\}}$ follow a Bernoulli mixture model with factor X; conditional default probabilities are given by $p_n(X) = \mathbb{P}(V_n \leq b_1^n | X)$.*

This relationship has already been observed for the CreditMetrics model and the CreditRisk$^+$ model in [70]. As Bernoulli mixture models can be more efficiently simulated, this proposition provides a method to effectively simulate certain types of latent variable models by using their Bernoulli mixture model analogues. We will make use of this property in Section 14.2 where we study the sensitivity of the tail of the portfolio loss variable on the dependence structure.

Example 5.3.3 (Multi-Factor Merton Model) Consider the multi-factor Merton model of Section 3.2 where the asset log-returns r_n of obligor n are given by equation (3.6), i.e.

$$r_n = \beta_n \cdot Y_n + \sqrt{1 - \beta_n^2} \cdot \varepsilon_n,$$

with composite factor

$$Y_n = \sum_{k=1}^{K} \alpha_{n,k} X_k$$

---

[2] Compare Definition 4.9 and Proposition 4.10 in [61].

5.3 Relation Between Latent Variable Models and Mixture Models

and standard normally distributed idiosyncratic shocks ε_n, independent of $X_1, \ldots, X_K$. The systematic risk factors $X_1, \ldots, X_K$ are also standard normally distributed. The latent variable model $(r_n, c_n)_{1 \leq n \leq N}$ is characterized by the default thresholds

$$c_n = \Phi^{-1}(\mathrm{PD}_n).$$

Then $r = (r_1, \ldots, r_N)$ has a K-dimensional conditional independence structure with factor vector $X = (X_1, \ldots, X_K)$. The equivalent Bernoulli mixture model is characterized by the default probabilities conditional on a realization x of the factor vector X which can be computed as (compare equation (3.8))

$$p_n(x) = \mathbb{P}(r_n \leq c_n | X) = \Phi\left(\frac{\Phi^{-1}(\mathrm{PD}_n) - \beta_n \sum_{k=1}^{K} \alpha_{n,k} x_k}{\sqrt{1 - \beta_n^2}} \right).$$

6
The CreditRisk$^+$ Model

In this chapter we will discuss the CreditRisk$^+$ model, another industry model which is based on a typical insurance mathematics approach and therefore, also often called an *actuarial model*. The CreditRisk$^+$ model has originally been developed by Credit Suisse Financial Products (CSFP) and is now one of the financial industry's benchmark models in the area of credit risk management. It is also widely used in the supervisory community since it uses as basic input the same data as also required for the Basel II IRB approach which we discussed in Chapter 4.

The CreditRisk$^+$ model represents a *reduced-form* model for portfolio credit risk. In contrast to the class of asset-value models, in these models the process of credit defaults is modeled directly instead of defining a stochastic process for the firm's asset value which then indirectly leads to defaults. The probabilities of default can vary depending on some underlying factors, hence default rates are not constant as in the KMV or CreditMetrics model but they are modeled as stochastic variables. Correlation between default events arises through dependence on some common underlying factors which influence the defaults of every single loan. It has the nice property that, instead of simulating the portfolio loss distribution, one can derive an analytic solution for the loss distribution of a given credit portfolio by means of an approximation of the probability generating function of the portfolio loss. Moreover, it seems easier to calibrate data to the model than it is the case for some multi-factor asset-value models. Most important, however, for this survey is that the CreditRisk$^+$ model reveals one of the most essential credit risk drivers: concentration.

We first explain the basic model setting and then proceed to the CreditRisk$^+$ model including the Poisson approximation approach. For a survey of the CreditRisk$^+$ model and its extensions we refer to [10].

6.1 Basic Model Setting

We consider the portfolio of N obligors introduced in Section 1.2. Obligor n constitutes a loss exposure E_n and has a probability of default PD_n over the time period $[0, T]$. The *loss exposure* is given by its exposure at default EAD_n times its expected percentage loss given default ELGD_n, i.e. $E_n = \text{EAD}_n \cdot \text{ELGD}_n$.

As in the previous models the state of obligor n at the time horizon T can be represented as a Bernoulli random variable D_n where

$$D_n = \begin{cases} 1 & \text{if obligor n defaults at time T,} \\ 0 & \text{otherwise.} \end{cases}$$

Hence the default probability is $\mathbb{P}(D_n = 1) = \text{PD}_n$ while the survival probability is given by $\mathbb{P}(D_n = 0) = 1 - \text{PD}_n$. In the full CreditRisk$^+$ model, which we will study in Subsection 6.3, the Bernoulli parameter PD_n is taken stochastic as well and the default indicator variables D_n are conditionally independent, i.e.

$$(D_n | \text{PD}_1, \ldots, \text{PD}_N)_n \text{ independent} \sim \text{Bern}(1; \text{PD}_n).$$

Definition 6.1.1
The *probability generating function* (PGF) of a non-negative integer-valued random variable X is defined as

$$G_X(z) = \mathbb{E}\left[z^X\right] = \sum_{i=0}^{\infty} z^i \cdot \mathbb{P}[X = i].$$

From this definition it immediately follows that

$$\mathbb{P}[X = i] = \frac{1}{i!} G_X^{(i)}(0), \quad i \in \mathbb{N}_0,$$

where $G_X^{(i)}(0)$ denotes the i-th derivative of the PGF $G_X(z)$ evaluated at $z = 0$. Thus the distribution of a random variable can easily be computed as soon as one knows the PGF. The CreditRisk$^+$ model makes use of exactly this property as well. Before returning to the derivation of the CreditRisk$^+$ model, we will briefly state some properties of the PGF which will be extensively used in the following.

Proposition 6.1.2 *Let X, Y be two random variables.*

1. *Let X, Y be independent. Then the PGF of $X + Y$ is simply the product of the PGF of X and the PGF of Y, i.e.*

$$G_{X+Y}(z) = G_X(z) \cdot G_Y(z).$$

2. Let $G_{X|Y}(z)$ be the PGF of X conditional on the random variable Y and denote the distribution function of Y by F. Then

$$G_X(z) = \int G_{X|Y=y}(z) F(dy).$$

Proof.

1. Since X, Y are independent, it follows that also z^X, z^Y are independent. Thus we have $\mathbb{E}[z^{X+Y}] = \mathbb{E}[z^X] \cdot \mathbb{E}[z^Y]$ which proves the first statement.
2. The second statement follows from

$$G_X(z) = \mathbb{E}[z^X] = \int \mathbb{E}[z^X | Y = y] F(dy) = \int G_{X|Y=y}(z) F(dy).$$

$\square$

Note that the probability generating function (PGF) of D_n can be computed to

$$G_{D_n}(z) = \sum_{x=0}^{\infty} \mathbb{P}(D_n = x) \cdot z^x = (1 - \mathrm{PD}_n) + \mathrm{PD}_n \cdot z. \qquad (6.1)$$

The loss of obligor n is given by the random variable $L_n = D_n \cdot E_n$. Hence we can compute the probability distribution of L_n as $\mathbb{P}(L_n = E_n) = \mathrm{PD}_n$ and $\mathbb{P}(L_n = 0) = 1 - \mathrm{PD}_n$. The total portfolio loss L is given by

$$L = \sum_{n=1}^{N} L_n = \sum_{n=1}^{N} D_n \cdot E_n.$$

In the CreditRisk$^+$ model the expected loss given defaults ELGD_n are usually modeled as constant fractions of the loan size. To limit the number of possible values for L the loss exposure amounts are expressed as integer multiplies of a fixed base unit of loss. Therefore one can, for example, choose the smallest exposure as the normalization factor and express all other exposures in multiples of this smallest exposure. Denote for example the base unit of loss by E, then the normalized exposure of obligor n is $\nu_n = EAD_n/E \cdot \mathrm{ELGD}_n = E_n/E$. In the following we denote the normalized loss of obligor n by $\lambda_n = D_n \cdot \nu_n$. Hence, λ_n is a random variable taking value ν_n with probability PD_n and value 0 with probability $1 - \mathrm{PD}_n$. The total normalized portfolio loss is $\lambda = \sum_{n=1}^{N} \lambda_n$.

In order to be able to compute the VaR for the portfolio (or equivalently any other risk measure for the portfolio) we need to derive the probability distribution of the portfolio loss L. Instead of determining the probability distribution of L, however, we can also derive the distribution of λ. For intuition, we will start with an example of a simplified version of the CreditRisk$^+$ model with non-random default probabilities.

Example 6.1.3 (Non-Random Default Probabilities) In the simplified case of non-random default probabilities, the PGF of the portfolio loss can easily be computed. Here we assume that individual default probabilities are known with certainty, and that default events are independent among obligors. The PGF of D_n is given by equation (6.1) from which we can easily derive the PGF of λ_n by the relation $\lambda_n = D_n \cdot \nu_n$. Since obligors are independent and since λ is the sum of λ_n over n, the PGF of λ is simply

$$G_\lambda(z) = \prod_{n=1}^{N} G_{\lambda_n}(z) = \prod_{n=1}^{N} \left[(1 - \text{PD}_n) + \text{PD}_n \cdot z^{\nu_n}\right].$$

6.2 The Poisson Approximation

When the default probabilities are random and default events are no longer independent, an analytical solution for the loss distribution can be derived by using an approximation for the distribution of the default events. Thus consider the individual default probabilities to be sufficiently small for the compound Bernoulli distribution of the default events to be well approximated by a Poisson distribution. Under this assumption it is still possible to derive an analytical solution of the loss distribution function.

Since default events are assumed to follow a Bernoulli distribution, the PGF of obligor n's default is given by equation (6.1), which can be reformulated as

$$G_{D_n}(z) = \exp[\ln(1 + \text{PD}_n \cdot (z - 1))]. \tag{6.2}$$

Assuming that PD_n is very small, $\text{PD}_n \cdot (z - 1)$ is also very small whenever $|z| \leq 1$. Applying a Taylor series expansion to $\ln(1 + w)$ around $w = 0$ we obtain

$$\ln(1 + w) = w - \frac{w^2}{2} + \frac{w^3}{3} \cdots.$$

Ignoring higher order terms we have, for $w = \text{PD}_n \cdot (z - 1)$,

$$\ln(1 + \text{PD}_n \cdot (z - 1)) \approx \text{PD}_n \cdot (z - 1).$$

Inserting this approximation in equation (6.2) we obtain

$$G_{D_n}(z) \approx \exp(\text{PD}_n \cdot (z - 1)).$$

Using again a Taylor series expansion of $\exp(\text{PD}_n \cdot (z - 1))$ around $z = 0$ we obtain

$$G_{D_n}(z) \approx \exp(\text{PD}_n \cdot (z - 1)) = \exp(-\text{PD}_n) \cdot \sum_{x=0}^{\infty} \frac{\text{PD}_n^x}{x!} z^x.$$

which is the PGF of the Poisson distribution with intensity PD_n. Hence for small values of PD_n, the Bernoulli distribution of D_n can be approximated by a Poisson distribution with intensity PD_n, i.e. we have approximately

$$\mathbb{P}(D_n = x) = \exp(-\mathrm{PD}_n) \cdot \frac{\mathrm{PD}_n^x}{x!}.$$

Moreover, the PGF of λ_n is defined as

$$G_{\lambda_n}(z) = \mathbb{E}\left[z^{\lambda_n}\right] = \mathbb{E}\left[z^{D_n \nu_n}\right]$$

which is equal to

$$\begin{aligned} G_{\lambda_n}(z) &= \sum_{x=0}^{\infty} \mathbb{P}(D_n = x) \cdot z^{D_n \nu_n} \\ &= \sum_{x=0}^{\infty} \exp(-\mathrm{PD}_n) \cdot \frac{\mathrm{PD}_n^x}{x!} \cdot (z^{\nu_n})^x \\ &= (1 + \mathrm{PD}_n\, z^{\nu_n}) \cdot \exp(-\mathrm{PD}_n) \end{aligned}$$

since D_n can only take the values 0 and 1. Due to Taylor expansion we have $\exp(\mathrm{PD}_n\, z^{\nu_n}) \approx 1 + \mathrm{PD}_n\, z^{\nu_n}$. Thus for small values of PD_n the PGF of λ_n can be approximated by

$$G_{\lambda_n}(z) = \exp\left[\mathrm{PD}_n \cdot (z^{\nu_n} - 1)\right]. \tag{6.3}$$

6.3 Model with Random Default Probabilities

Up to now we have not used any particular assumptions of the CreditRisk+ model. However, to obtain an analytic solution for the loss distribution we have to impose some assumptions on the model.

Assumption 6.3.1
Assume that the default probabilities are random and that they are influenced by a common set of Gamma-distributed systematic risk factors. Thus, the default events are assumed to be mutually independent only conditional on the realizations of the risk factors.

These are the original assumptions of the CreditRisk+ model. The above Poisson approximation can still be applied and allows us to obtain, even in this setting, an analytical solution for the loss distribution.

As already mentioned, in the CreditRisk+ model, correlation among default events is induced by the dependence of the default probabilities on a common set of risk factors. Assume that there exist K systematic risk factors $X_1, \ldots, X_K$, indexed by $k = 1, \ldots, K$, which describe the variability of the default probabilities PD_n. Each factor is associated with a certain *sector*, for example an industry, country or region. All risk factors are taken to be independent and Gamma distributed with shape parameter $\alpha_k = 1/\xi_k$ and scale parameter $\beta_k = \xi_k$. Recall that the Gamma distribution is defined by the probability density

$$\Gamma_{\alpha,\beta}(x) := \frac{1}{\beta^\alpha \Gamma(\alpha)} \cdot e^{-x/\beta} \cdot x^{\alpha-1},$$

for $x \geq 0$, where $\Gamma(\cdot)$ denotes the Gamma function. The density functions of two Gamma distributions are plotted in Figure 6.1. The black graph corresponds to a Gamma distribution with shape parameter $1/8$ and scale parameter 8, whereas the grey graph corresponds to a $\Gamma_{2,1/2}$ distribution. The mean and variance of X_k can be computed to

$$\mathbb{E}[X_k] = \alpha_k \cdot \beta_k = 1 \text{ and } \mathbb{V}[X_k] = \alpha_k \cdot \beta_k^2 = \xi_k.$$

Fig. 6.1. Densities of Gamma distributions

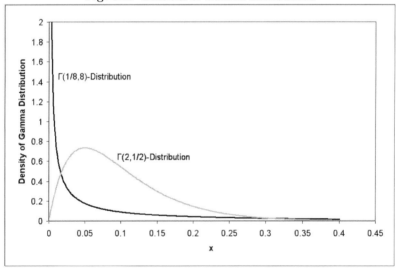

Definition 6.3.2
The *moment generating function* (MGF) of a random variable Y with density f_Y is defined as the analytic function

$$M_Y(z) = \mathbb{E}\left[e^{zY}\right] = \int e^{zt} f_Y(t) dt$$

of a complex variable z, provided that the integral exists.

6.3 Model with Random Default Probabilities

By this definition the MGF of X_k can be computed as

$$M_{X_k}(z) = \mathbb{E}\left[\exp\left(X_k \cdot z\right)\right] = (1 - \beta_k \cdot z)^{-\alpha_k} = (1 - \xi_k \cdot z)^{-1/\xi_k}.$$

For simplicity of notation we denote the idiosyncratic risk by $X_0 = 1$ and let $X = (X_0, X_1, \ldots, X_K)$. The link between the default probabilities and the risk factors X_k is given by the following factor model

$$\text{PD}_n(X) = \text{PD}_n\left(\sum_{k=0}^{K} w_{k,n} X_k\right), \quad n = 1, \ldots, N, \tag{6.4}$$

where PD_n is the average default probability of obligor n, and the factor loading $w_{k,n}$ for $k = 1, \ldots, K$, measures the sensitivity of obligor n to the risk factor X_k, where $0 \leq w_{k,n} \leq 1$ and for all $n = 1, \ldots, N$,

$$\sum_{k=1}^{K} w_{k,n} \leq 1.$$

Define $w_{0,n} = 1 - \sum_{k=1}^{K} w_{k,n}$ as the share of idiosyncratic risk of obligor n.

As before, D_n is assumed to follow a Poisson distribution with intensity $\text{PD}_n(X)$. The PGF of the individual normalized loss λ_n is assumed to be given by equation (6.3), which by the above factor model specification can be rewritten as

$$G_{\lambda_n}(z) = \exp\left[\text{PD}_n \cdot \left(\sum_{k=0}^{K} w_{k,n} X_k\right) \cdot (z^{\nu_n} - 1)\right]$$

$$= \prod_{k=0}^{K} \exp\left[\text{PD}_n \cdot w_{k,n} \cdot X_k \cdot (z^{\nu_n} - 1)\right].$$

We denote by $G_\lambda(z|X)$ the PGF of the total normalized loss, conditional on X. Since individual losses are mutually independent conditional on X, the PGF of the normalized portfolio loss $G_\lambda(z|X)$ is the product of the individual loss PGFs conditional on X. Thus we have

$$G_\lambda(z|X) = \prod_{n=1}^{N} G_{\lambda_n}(z|X)$$

$$= \prod_{n=1}^{N} \prod_{k=0}^{K} \exp\left[\text{PD}_n \cdot w_{k,n} \cdot X_k \cdot (z^{\nu_n} - 1)\right]$$

$$= \exp\left[\sum_{n=1}^{N} \sum_{k=0}^{K} \text{PD}_n \cdot w_{k,n} \cdot X_k \cdot (z^{\nu_n} - 1)\right].$$

Defining $P_k(z) := \sum_{n=1}^{N} \text{PD}_n \cdot w_{k,n} \cdot (z^{\nu_n} - 1)$, we can rewrite $G_\lambda(z|X)$ as

$$G_\lambda(z|X) = \exp\left[\sum_{k=0}^{K} X_k \cdot P_k(z)\right].$$

We denote by $\mathbb{E}_X$ the expectation operator under the probability distribution of X. The unconditional PGF of the total normalized loss, denoted by $G_\lambda(z)$, is the expectation of $G_\lambda(z|X)$ under X's probability distribution, i.e.

$$G_\lambda(z) = \mathbb{E}_X\left[G_\lambda(z|X)\right] = \mathbb{E}_X\left[\exp\left(\sum_{k=0}^{K} X_k \cdot P_k(z)\right)\right].$$

For $P(z) = (P_0(z), \ldots, P_K(z))$ and $X = (X_0, \ldots, X_K)$ this is by definition of the joint MGF M_X of X equal to

$$G_\lambda(z) = \mathbb{E}_X\left[\exp\left(P(z) \cdot X\right)\right] = M_X\left[P(z)\right].$$

Due to the independence of the variables X_k this can be rewritten as

$$G_\lambda(z) = M_X\left[P(z)\right]$$

$$= \prod_{k=0}^{K} M_{X_k}\left[P_k(z)\right]$$

$$= \prod_{k=0}^{K} \left[1 - \xi_k \cdot P_k(z)\right]^{-1/\xi_k}$$

$$= \prod_{k=0}^{K} \exp\left(\ln\left(\left[1 - \xi_k \cdot P_k(z)\right]^{-1/\xi_k}\right)\right)$$

$$= \exp\left(-\sum_{k=0}^{K} \frac{1}{\xi_k} \ln\left(1 - \xi_k \cdot P_k(z)\right)\right) =: G_{\lambda,CR+}(z)$$

which is the PGF of the total normalized portfolio loss for the CreditRisk$^+$ model.

Part II

Concentration Risk in Credit Portfolios

7
Introduction

When measuring credit risk, we are particularly interested in dependencies between certain extreme credit events. In connection with the LTCM case the *Business Week* stated in September 1998:

Extreme, synchronized rises and falls in financial markets occur infrequently but they do occur. The problem with the models is that they did not assign a high enough chance of occurrence to the scenario in which many things go wrong at the same time – the "perfect storm" scenario.

This perfect storm scenario is what we mean by concentration of risk. The quote underlines the importance of a sufficient measurement of concentration risk since losses due to concentration risk can be extreme.

Historical experience shows that concentration of risk in asset portfolios has been one of the major causes of bank distress. The failures of large borrowers like Enron, Worldcom and Parmalat were the source of sizeable losses in a number of banks. In the 1980's banks in Texas and Oklahoma suffered severe losses in both corporate and commercial real estate lending due to significant concentrations of lending in the energy industry. Moreover the regional dependence on oil caused strong correlation between the health of the energy industry and local demand for commercial real estate. Another well known example is the insolvency of the German Schmidt Bank in 2001. The institution was highly concentrated in a less developed region with a fragile and concentrated industry structure.

Furthermore, the relevance of sector concentration risk is demonstrated by the recent developments in conjunction with the *subprime mortgage crisis*, which started in the United States in late 2006 and turned into a global financial crisis during 2007 and 2008. Subprime refers to loans granted to borrowers of low credit-worthiness. In the United States many borrowers speculated on rising housing prices and assumed mortgages with the intention to profitably refinance them later on. However, housing prices started to drop in 2006 and 2007 so that refinancing became increasingly difficult and, finally, lead to several defaults. Many mortgage lenders had used securitization methods as, for example, mortgage-backed securities (MBS) or collateralized debt obligations (CDO). Thereby, the credit risk associated with subprime lending had been distributed broadly to corporate, individual and institutional investors holding these MBSs and CDOs. High mortgage payment defaults then lead to significant losses for financial institutions, accumulating to US$170 billion in March 2008. The crisis had spread out worldwide. In Germany, for example, the IKB Deutsche Industriebank also suffered dramatic losses from the subprime market downturn, leading to a bail out in August 2007. On August 13, 2007, the *MarketWatch* cited Chris Brendler and Michael Widner, economic analysts at Stifel Nicolaus & Co, claiming:

The rapidly increasing scope and depth of the problems in the mortgage market suggest that the entire sector has plunged into a downward spiral similar to the subprime woes whereby each negative development feeds further deterioration.

This also points out the importance of a proper measurement and management of concentration risk. Within the subprime or mortgage sector, loans are highly correlated and defaults spread quickly by contagion effects. In January 2008, *The Banker* published an analysis by Jacques de Larosiére, the former chairman of the International Monetary Fund (IMF), on the subprime crisis. He suggests as an issue, that needs to be addressed more seriously in future:

Risk concentration is another issue that must be addressed. Is it normal that some financial institutions have engineered SIVs [structured investment vehicles] with assets amounting to several times their capital? The off-balance-sheet character of these conduits seems to have encouraged some institutions to disregard the wise old rule of 'never put more than 25% of your equity on one client'. It is all the more relevant that many SIVs are concentrated in one sector (for instance, mortgage or subprime). In this case also, the reform of capital adequacy constraints partly addresses the issue (Pillar II).

In addition to these examples, the amount of specific rules, that are imposed by the supervisory authorities to control concentration of risks, shows the importance of diversification of loan portfolios with respect to regions, countries and industries. Think for example of the *granularity adjustment* introduced in the *Second Consultative Paper* [12] which aimed at an adjustment of the IRB model for concentrations in single-name exposures. Although it had been abandoned later on in the *Third Consultative Paper* [13], it was recently recovered by supervisory authorities for example in [41] and [9]. These examples, rules and ongoing research demonstrate that it is substantial for the evaluation and management of a bank's credit risk to identify and measure concentration of risks within a portfolio.

It is the aim of these lecture notes to reflect the recent developments in research on concentration risk both from an academic and from a supervisory perspective. To fix terminology we start here in this introductory chapter with the definition of concentration risk in loan portfolios on which the fur-

7 Introduction   65

ther discussions will be based. Of course there are various ways to define concentration, however, from the perspective of credit risk management we consider the following as the most appropriate one.

Definition 7.0.3 (Concentration Risk)
Concentration risks in credit portfolios arise from an unequal distribution of loans to single borrowers (*name concentration*) or different industry or regional sectors (*sector* or *country concentration*). Moreover, certain dependencies as, for example, direct business links between different borrowers, or indirectly through credit risk mitigation, can increase the credit risk in a portfolio since the default of one borrower can cause the default of a dependent second borrower. This effect is called *default contagion* and is linked to both name and sector concentration.

We will discuss the topic of default contagion in Part III of these lecture notes. Related to this definition of concentration risk are of course a number of further questions: what do we mean by a *sector* and how would we define this? One could define a sector in a way that each obligor in the sector should be an equally good substitute for each other obligor within the same sector. However, this definition is quite far away from possible fulfillment in practice. So what would be an applicable and sensible definition of a sector?

What is meant by an unequal distribution of loans? Do we mean this in terms of exposure sizes in absolute or in relative sense? Moreover, do we mean by a loan the entity of exposure size, default probability, correlations with other loans in the portfolio, etc.? Of course a model taking into account all of these influencing components will not be applicable in practice as data availability will determine the limits of a model. Thus we will have to judge which variables are the most important ones. In the following chapters some of these questions will be answered for particular models and we will discuss the advantages and shortcomings of the individual approaches. However, before turning to the modeling of concentration risk let us come back once more to the definition above and the relation to other sources of credit risk.

In the portfolio risk-factor frameworks that underpin both industry models of credit VaR and the Internal Ratings Based (IRB) risk weights of Basel II, credit risk in a portfolio arises from two sources, systematic and idiosyncratic:

- *Systematic risk* represents the effect of unexpected changes in macroeconomic and financial market conditions on the performance of borrowers. Borrowers may differ in their degree of sensitivity to systematic risk, but few firms are completely indifferent to the wider economic conditions in which they operate. Therefore, the systematic component of portfolio risk is unavoidable and only partly diversifiable.
- *Idiosyncratic risk* represents the effects of risks that are particular to individual borrowers. As a portfolio becomes more fine-grained, in the sense that the largest individual exposures account for a smaller share of total

portfolio exposure, idiosyncratic risk is diversified away at the portfolio level.

Let us explain the issue of concentration risk in more detail based on the example of the Asymptotic Single Risk Factor (ASRF) model. Under the ASRF framework that underpins the IRB approach, it is assumed that bank portfolios are *perfectly* fine-grained, that is, that idiosyncratic risk has been fully diversified away, so that economic capital depends only on systematic risk. Real-world portfolios are not, of course, perfectly fine-grained. The asymptotic assumption might be approximately valid for some of the largest bank portfolios, but clearly would be much less satisfactory for portfolios of smaller or more specialized institutions. When there are material name concentrations of exposure, there will be a residual of undiversified idiosyncratic risk in the portfolio, and the IRB formula will understate the required economic capital. Moreover, the single risk factor assumption of the ASRF model does not allow for the explicit measurement of sector concentration risk. However, a portfolio consisting only of exposures to the energy industry will be much more risky than a portfolio with exposures diversified over several different industrial sectors. Finally, let us mention that both components of credit risk cannot be discussed independently in the light of concentration risk. A portfolio which can be considered as almost perfectly granular and therefore non-concentrated in the sense of single-name concentration might be highly concentrated in the sense of sector concentration due to default dependencies between the loans in the portfolio.

Similar problems occur in most of the industry models for measuring credit risk as well. Therefore, we will discuss in the following chapters various different approaches for measuring concentration risks in credit portfolios. These methods will sometimes be based on one or the other industry model as well as on the ASRF model underpinning the IRB approach of Basel II. Prior to the discussion of the specific approaches we will briefly describe some general ad-hoc measures which can be applied to any variation of concentration risk but which also have severe shortcomings.

8
Ad-Hoc Measures of Concentration

Before we start to discuss different model-based methods to measure concentration risk in credit portfolios in the following chapters, we will first give a brief overview of some simple ad-hoc measures for concentration risk. We will discuss the advantages and drawbacks of these measures based on a set of desirable properties which ensure a consistent measurement of concentration risk. These properties are taken from [53] and [78], originally coming from the theory of concentrations in industry. A detailed study of ad-hoc measures based on these properties can be found in [17] who also translated them to the context of concentrations in credit portfolios.

A concentration index for a portfolio of N loans should satisfy the following properties (compare [17]):

(1) The reduction of a loan exposure and an equal increase of a bigger loan must not decrease the concentration measure (*transfer principle*).
(2) The measure of concentration attains its minimum value, when all loans are of equal size (*uniform distribution principle*).
(3) If two portfolios, which are composed of the same number of loans, satisfy that the aggregate size of the k biggest loans of the first portfolio is greater or equal to the size of the k biggest loans in the second portfolio for $1 \leq k \leq N$, then the same inequality must hold between the measures of concentration in the two portfolios (*Lorenz-criterion*).
(4) If two or more loans are merged, the measure of concentration must not decrease (*superadditivity*).
(5) Consider a portfolio consisting of loans of equal size. The measure of concentration must not increase with an increase in the number of loans (*independence of loan quantity*).
(6) Granting an additional loan of a relatively low amount does not increase the concentration measure. More formally, if $\bar{s}$ denotes a certain percentage of the total exposure and a new loan with a relative share of $s_n \leq \bar{s}$ of the total exposure is granted, then the concentration measure does not increase (*irrelevance of small exposures*).

All these properties are essential for an index to qualify as a measure of concentration. More properties which are desirable for analytical convenience can be found in [78]. However, as they are not essential for a concentration index we omit the description here. [53] proved that property (1) implies properties (2) and (3) while property (5) follows from properties (2) and (4). Furthermore, [78] showed that properties (1) and (6) imply property (4). Here the argument is the following: a merger can be decomposed in two stages. First we transfer most of the smaller exposure to the bigger one. Due to the transfer principle (1) we know that this must not decrease the concentration measure. Then we remove the remaining small exposure. Due to the irrelevance of small exposures principle (6) this does not decrease the concentration measure as well. Hence a merger cannot decrease the concentration measure. Thus if an index satisfies properties (1) and (6) then it satisfies all six properties.

This list of properties now enables us to discuss the applicability and information content of individual indices used in the literature on concentration risk.

8.1 Concentration Indices

The probably simplest method to quantify concentration in a portfolio is to compute the share of exposure held by the k largest loans in the portfolio relative to the total portfolio exposure for some $1 \leq k \leq N$. Here the total exposure is given as the sum over all individual exposures. The notion of a concentration ratio is based on exactly this idea.

Definition 8.1.1 (Concentration Ratio)
Consider a portfolio with exposure shares $s_1 \geq s_2 \geq \ldots \geq s_N$ and such that $\sum_{n=1}^{N} s_n = 1$. Then the *Concentration Ratio* (CR_k) of the portfolio is defined as the ratio of the sum of the k biggest exposures to the total sum of exposures in the portfolio

$$CR_k(s_1, \ldots, s_N) = \sum_{i=1}^{k} s_i \quad \text{for} \quad 1 \leq k \leq N.$$

The concentration ratio satisfies all six properties as can easily be seen. However, it has quite a number of deficiencies as a measure of concentration. The number k is chosen by the investigator or the risk manager and, therefore, the choice of k is arbitrary although it has a strong impact on the outcome. Moreover, the ratio considers only the size distribution of the k largest loans and does not take into account the full information about the loan distribution. Furthermore, shifts in the portfolio structure can stay unrecognized by this measure depending on the choice of k. One can circumvent this problem by considering all possible concentration ratios, i.e. by constructing a *concentration curve* where the number k of loans taken into account in the computation of the concentration ratio is mapped against the corresponding concentration ratio. If the concentration curve of a portfolio A lies entirely below that of another portfolio B then it seems natural to state that B is more concentrated than A. However, frequently concentration curves intersect such that we cannot give a proper ranking of the concentration level of different portfolios.

Closely related to the concentration curve is the *Lorenz curve* (see [96]). Similarly to the concentration curve, the Lorenz curve is not exactly an *index* in the sense that it returns a single number for each credit portfolio. It is a mapping that assigns to every percentage q of the total loan number the cumulative percentages $L(q)$ of loan sizes. This is formalized in the following definition (see [65]).

Definition 8.1.2 (Lorenz Curve)
Given an ordered data set $A_1 \leq A_2 \leq \ldots \leq A_N$, the *(empirical) Lorenz curve* generated by the data set is defined for all $q \in (0,1)$ as the piecewise linear interpolation with breakpoints $L(0) = 0$ and

$$L(n/N) = \frac{\sum_{i=1}^{n} A_i}{\sum_{j=1}^{N} A_j}, \quad n = 1, \ldots, N.$$

Consider a distribution function F with density function f such that F increases on its support and that the mean of F exists. Then the qth quantile $F^{-1}(q)$ of F is well defined and the *(theoretical) Lorenz curve* $L(q)$ for $q \in (0,1)$ is defined by

$$L(q) = \frac{\int_0^q F^{-1}(t)dt}{\int_{-\infty}^{+\infty} tf(t)dt}.$$

Reformulating, we obtain by substitution

$$L(F(x)) = \frac{\int_{-\infty}^{F(x)} F^{-1}(t)dt}{\int_{-\infty}^{+\infty} tf(t)dt} = \frac{\int_{-\infty}^{x} tf(t)dt}{\int_{-\infty}^{+\infty} tf(t)dt}.$$

Hence the Lorenz curve is a graph showing for each level q the proportion of the distribution assumed by the first q percentage of values. Obviously, a Lorenz curve always starts at $(0,0)$ and ends at $(1,1)$ and is by definition a continuous function. If the variable being measured cannot take negative values, the Lorenz curve is an increasing and convex function.

The line of perfect equality in the Lorenz curve is $L(F(x)) = F(x)$ and represents a uniform distribution while perfect inequality in the Lorenz curve is characterized by $L(F(x)) = \delta(x)$, the Dirac function with weight 1 at $x = 1$, representing a Dirac distribution. The dotted graph in Figure 8.1 shows the Lorenz curve for a portfolio of $\Gamma(3,1)$-distributed loans. The solid graph

Fig. 8.1. Lorenz curve for a $\Gamma(1,3)$ distribution

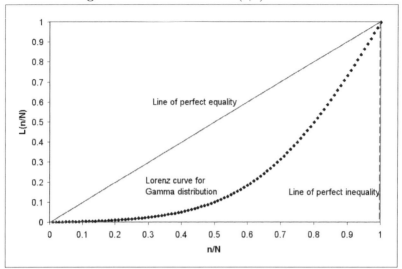

corresponds to the line of perfect equality while the dashed graph represents perfect inequality.

Similar to the concentration curve, the main drawback of the Lorenz curve as a measure of concentration is that it does not allow a unique ranking of two portfolios in terms of their concentration. The latter is only possible when the curves of the two portfolios do not intersect. In that case the portfolio with the lower Lorenz curve is said to be higher concentrated. Although we cannot prove the properties stated above, one can in general say that the Lorenz curve is not optimal as a measure for concentration risk since it does not deliver a unique ranking of loans and rather measures a deviation from the uniform distribution where the number of loans in the portfolio is not taken into account. We want to point out here that inequality and concentration are, of course, not the same since concentration also depends on the number of loans in the portfolio. A portfolio of two loans of the same size will be considered as well-diversified when applying the Lorenz curve as a concentration measure. However, a portfolio consisting of one hundred loans of different sizes might be considered as much more concentrated as its Lorenz curve will differ from the line of perfect equality. Thus, it might be questionable to measure the degree of concentration on the basis of deviation from equality.

Another ad-hoc measure which is closely linked to the Lorenz curve is the *Gini Coefficient* which measures the size of the area between the Lorenz curve and the main diagonal and thus also represents a measure for the deviation from equal distribution.

8.1 Concentration Indices

Definition 8.1.3 (Gini Coefficient)
For a portfolio of N loans with exposure shares $s_1, \ldots, s_N$, the *(empirical) Gini coefficient* is defined as

$$G(s_1, \ldots, s_N) = \frac{\sum_{n=1}^{N}(2n-1)s_n}{N} - 1.$$

For a given distribution function the *(theoretical) Gini coefficient* is defined as the ratio of the area between the Lorenz curve $L(q)$ of the distribution and the curve of the uniform distribution, to the area under the uniform distribution.

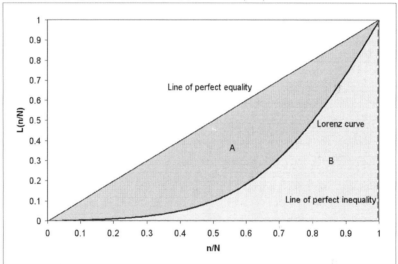

Fig. 8.2. Gini coefficient for a $\Gamma(1,3)$ distribution

If the area between the line of perfect equality and the Lorenz curve is denoted by A, and the area under the Lorenz curve is denoted by B, then the Gini coefficient is $G = A/(A+B)$. Figure 8.2 illustrates the relation between the Gini coefficient and the Lorenz curve for a portfolio with $\Gamma(1,3)$-distributed loans. Since $A+B=0.5$, the Gini coefficient equals $G = 2A = 1 - 2B$. For the portfolio with Gamma distributed exposures, the Gini coefficient equals approximately 0.52. If the Lorenz curve is represented by the function $L(q)$, the value of B can be found with integration and

$$G = 1 - 2\int_0^1 L(q)dq.$$

A value for the Gini coefficient close to zero corresponds to a well diversified portfolio where all exposures are more or less equally distributed and a value close to one corresponds to a highly concentrated portfolio.

One can easily see that properties (1) and (5) are satisfied by the Gini coefficient. However, properties (4) and (6) are violated as shown in [17]. Similar to the Lorenz curve, a major drawback of the Gini coefficient for the measurement of concentration risks, however, is the fact that the portfolio size is not taken into account. Moreover, the Gini coefficient can increase when a relatively small loan is added to a given portfolio although concentration would actually decrease. Therefore, we consider the Gini coefficient suitable to only a limited extend for the measurement of concentration risks.

The most commonly used *ad-hoc* or model-free measure is probably the so-called *Herfindahl-Hirschman Index* (HHI). It is used extensively particularly in the empirical literature.

Definition 8.1.4 (Herfindahl-Hirschman Index)
The *Herfindahl-Hirschman Index* (HHI) is defined as the sum of squared market shares (measured in fractions of the total portfolio) of each market participant, i.e.

$$\text{HHI} = \frac{\sum_{n=1}^{N} A_n^2}{\left(\sum_{n=1}^{N} A_n\right)^2} = \sum_{n=1}^{N} s_n^2.$$

The HHI is a continuous measure with zero corresponding to the fully granular case (each participant has an infinitesimal share) and unity corresponding to monopoly (there is only one participant). Holding all else equal, the closer the HHI of a portfolio is to 1, the more concentrated the portfolio is, and hence the higher the appropriate granularity add-on charge. It can be shown (see [17]) that the Herfindahl-Hirschman Index satisfies all six properties.

Note that for the portfolio with $\Gamma(1,3)$-distributed loans, we already used before for illustrative purposes, the HHI would approximately equal 0.019. Thus, the portfolio could be considered as quite well diversified when using the HHI although its Gini coefficient is of average level with a value of 0.52.

8.2 Conclusion

For any of the discussed ad-hoc approaches it is difficult to say what the "appropriate" add-on for a given index value should be since they do not give an explicit capital charge which should be kept for concentration risk.
Moreover, neither the HHI nor the Gini coefficient or any other model-free method for measuring concentration risk can incorporate the effects of obligor specific credit qualities, which are, for example, represented by obligor-specific

default probabilities. For these reasons certain *model-based* methods for measuring concentration risks have been developed which can deal more explicitly with exposure distribution, credit quality, and default dependencies. In the following chapters we will discuss some of these model-based approaches for the measurement of concentration risk.

9
Name Concentration

This chapter focuses on the measurement of single-name concentrations in loan portfolios. As already mentioned in Chapter 7, we can decompose credit risk in loan portfolios into a systematic and an idiosyncratic component. Systematic risk represents the effect of unexpected changes in macroeconomic and financial market conditions on the performance of borrowers while idiosyncratic risk represents the effects of risks that are particular to individual borrowers.

When applying a single-factor credit portfolio model such as the one-factor Vasicek model, the portfolio loss variable can be written as a function of the risk factor representing the systematic risk and a residual characterizing the idiosyncratic risk. To determine explicitly the contribution of the idiosyncratic component to credit Value-at-Risk, a standard method is to apply Monte Carlo simulations. This can be computationally quite time consuming, especially because we are interested in events far in the tail of the loss distribution and, therefore, we need a very large number of trials to obtain an accurate result. In this chapter we present some analytical alternatives. As a portfolio becomes more fine-grained, in the sense that the largest individual exposures account for a smaller share of total portfolio exposure, idiosyncratic risk is diversified away at the portfolio level. Under some conditions it can then be shown that the portfolio loss variable converges to a function which is solely determined by the systematic risk factor. We will call such a portfolio *infinitely fine-grained*. Since real-world portfolios are not, of course, perfectly fine-grained, there will be a residual of undiversified idiosyncratic risk in every portfolio. The impact of undiversified idiosyncratic risk on portfolio VaR can be approximated analytically and is known as a *granularity adjustment* (GA). This idea has been introduced in 2000 by [72] for application in Basel II. It was then substantially refined and put on a more rigorous foundation by [122] and [102].[1]

---

[1] The results of [102] can be viewed as an application of theoretical work by [76]. Other early contributions to the GA literature include [121] and [110]. A nice

76   9 Name Concentration

In Section 9.1 we will present the theoretical derivation of such a GA based on the recent work [75]. In Section 9.2 we present an alternative approach to capital charges by [52] which relies on a limiting procedure applied to a part of the portfolio only and which is therefore called the *semi-asymptotic approach*. [52] present a second method which is quite close in spirit and methodology to the GA approach of [75]. They offer a granularity adjustment based on a one-factor default-mode CreditMetrics model, which has the advantage of relative proximity of the model underpinning the IRB formula.

Finally, an alternative that has not been much studied in the literature is the saddle-point based method of [102]. Results in that paper suggest that it would be quite similar to the GA in performance and pose a similar tradeoff between fidelity to the IRB model and analytical tractability. Indeed, it is not at all likely that the saddle-point GA would yield a closed-form solution for any industry credit risk model other than CreditRisk$^+$. We will discuss this approach in Section 9.3. Section 9.4 comprises a comparison study of the different methods to measure name concentration risk and a detailed discussion of certain regulatory problems that might arise within the GA approach.

Before introducing the different methodologies for the measurement of name concentration, we want to point out here, right in the beginning, that it is advisable to measure name concentration or granularity on obligor level and not on exposure level. Measuring name concentration based on the single exposures in the portfolio can lead to an underestimation of the real concentration risk since the risk consists in the default of certain obligors and not in the default of single exposures. Being aware of this, one of the main problems in practice is indeed the aggregation of exposures in the sense that all exposures to a single borrower have to be aggregated before applying any method to quantify name concentration. This problem is common to any method, whether ad-hoc or model-based, for measuring name concentration.

9.1 A Granularity Adjustment for the ASRF Model

The Asymptotic Single Risk Factor (ASRF) model discussed in Chapter 4 presumes that portfolios are fully diversified with respect to individual borrowers, so that economic capital depends only on systematic risk. Hence, the IRB formula omits the contribution of the residual idiosyncratic risk to the required economic capital. This form of credit concentration risk is sometimes also called *lack of granularity*. In this section we discuss an approach how to assess a potential add-on to capital for the effect of lack of granularity first in a general risk-factor framework and then, in particular, in the ASRF model presented in Chapter 4. In our presentation of the *granularity adjustment*

---

survey of these developments and a primer on the mathematical derivation is presented in [73].

9.1 A Granularity Adjustment for the ASRF Model

(GA) we follow the *revised* methodology developed in [75] which is similar in form and spirit to the granularity adjustment that was included in the Second Consultative Paper (CP2) [12] of Basel II. Like the CP2 version, the revised GA is derived as a first-order asymptotic approximation for the effect of diversification in large portfolios within the CreditRisk$^+$ model of portfolio credit risk. Also in keeping with the CP2 version, the data inputs to the revised GA are drawn from quantities already required for the calculation of IRB capital charges and reserve requirements. However, it takes advantage of theoretical advances that have been made since the time of CP2 and the resulting algorithm is both simpler and more accurate than the one of CP2.

9.1.1 Example as Motivation for GA Methodology

Before we introduce the general framework for the granularity adjustment, let us give a simple and intuitive example of how the GA methodology works. Assume that the single systematic risk factor X is normally distributed with mean 0 and variance ν^2 and that the loss rate on instrument n, conditional on the systematic risk factor $U_n|X$, is also normally distributed with mean X and variance σ^2, where ν, σ are some known constants. Then the unconditional distribution of the portfolio loss ratio L_N is also normally distributed. From this distribution we can compute the qth quantile which is given by $\sqrt{\nu^2 + \sigma^2/N} \cdot \Phi^{-1}(q)$ where Φ denotes the standard normal cumulative distribution function. As the number N of obligors in the portfolio increases to infinity, the distribution of L_N converges to that of X. Thus the quantile of the asymptotic distribution is $\nu \cdot \Phi^{-1}(q)$. This is the systematic component of VaR. The idiosyncratic component can be derived as the difference between the quantile of the actual loss ratio L_N, which incorporates both the systematic and the idiosyncratic component, and the asymptotic loss L_∞, which represents the systematic part only.

It can be shown by a simple application of Taylor's expansion around $\sigma^2/N = 0$ that $\sqrt{\nu^2 + \sigma^2/N}$ can be approximated by $\nu + \frac{1}{N}\frac{\sigma^2}{2\nu}$ plus some terms proportional to $1/N^2$. Thus the idiosyncratic component can be written as $\frac{1}{N}\frac{\sigma^2}{2\nu}\Phi^{-1}(q) + \mathcal{O}(N^{-2})$. The GA is an application of exactly the same logic to a proper credit risk model, in the case of [75] the single factor CreditRisk$^+$ model.

Assume that the systematic risk factor is standard normal, hence ν in the example is assumed to be 1. Figure 9.1 shows the systematic and idiosyncratic components of VaR of the portfolio loss ratio L_N as the number N of obligors in the portfolio increases. The dotted line shows the actual VaR (hence both the systematic and the idiosyncratic component) and the dashed line shows the systematic component only. As the number N of borrowers in the portfolio increases the idiosyncratic component vanishes. This is also the main intuition of the GA. For large portfolios, which are typically better diversified, the GA is lower than for small concentrated portfolios.

Fig. 9.1. Systematic and idiosyncratic component of VaR of portfolio loss L_N

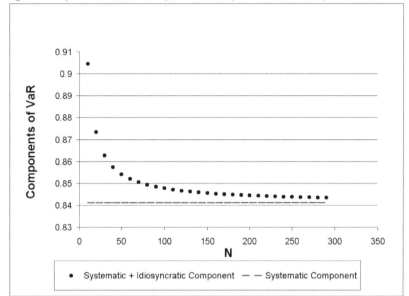

9.1.2 The General Framework

The GA has been developed as an extension of the ASRF model which underpins the IRB model. In principle, however, it can be applied to any risk-factor model for portfolio credit risk, and so we begin with a general framework.

Let X denote the systematic risk factor. For simplicity of notation and for consistency with the ASRF framework we assume that X is unidimensional, i.e. there is only a single systematic risk factor. For our portfolio of N risky loans described in Chapter 1, let U_n denote the loss rate on position n. Let L_N be the loss rate on the portfolio of the first N positions, i.e.

$$L_N = \sum_{n=1}^{N} s_n \cdot U_n. \tag{9.1}$$

When economic capital is measured as Value-at-Risk at the qth percentile, we wish to estimate $\alpha_q(L_N)$. The IRB formula, however, delivers the qth percentile of the expected loss conditional on the systematic factor $\alpha_q(\mathbb{E}[L_N|X])$. The difference $\alpha_q(L_N) - \alpha_q(\mathbb{E}[L_N|X])$ is the "exact" adjustment for the effect of undiversified idiosyncratic risk in the portfolio. Such an exact adjustment cannot be obtained in analytical form, but we can construct a Taylor series approximation in orders of $1/N$. Define the functions

9.1 A Granularity Adjustment for the ASRF Model

$$\mu(X) = \mathbb{E}[L_N|X] \quad \text{and} \quad \sigma^2(X) = \mathbb{V}[L_N|X]$$

as the conditional mean and variance of the portfolio loss, respectively, and let h be the probability density function of X.

For $\varepsilon = 1$ the portfolio loss L_N is given by

$$L_N = \mathbb{E}[L_N|X] + \varepsilon(L_N - \mathbb{E}[L_N|X]). \tag{9.2}$$

The asymptotic expansion for the calculation of the qth quantile of the portfolio loss as shown in [121] is based on a 2nd order Taylor expansion in powers of ε around the conditional mean $\mathbb{E}[L_N|X]$. Evaluating the resulting formula at $\varepsilon = 0$, we obtain

$$\alpha_q(L_N) = \alpha_q(\mathbb{E}[L_N|X]) + \frac{\partial}{\partial \varepsilon}\alpha_q(\mathbb{E}[L_N|X] + \varepsilon \cdot (L_N - \mathbb{E}[L_N|X]))|_{\varepsilon=0} \tag{9.3}$$
$$+ \frac{1}{2}\frac{\partial^2}{\partial \varepsilon^2}\alpha_q(\mathbb{E}[L_N|X] + \varepsilon \cdot (L_N - \mathbb{E}[L_N|X]))|_{\varepsilon=0} + O(\varepsilon^3).$$

Hence the granularity adjustment of the portfolio is given by

$$GA_N = \frac{\partial}{\partial \varepsilon}\alpha_q(\mu(X) + \varepsilon \cdot (L_N - \mu(X)))|_{\varepsilon=0} \tag{9.4}$$
$$+ \frac{1}{2}\frac{\partial^2}{\partial \varepsilon^2}\alpha_q(\mu(X) + \varepsilon \cdot (L_N - \mu(X)))|_{\varepsilon=0}.$$

The following Lemma is due to [76], Lemma 1. We will use this result to prove that the first derivative in the Taylor expansion of the quantile vanishes.

Lemma 9.1.1 *Consider a bivariate continuous vector (X,Y) and the qth quantile $Q(q, \varepsilon) := \alpha_q(X + \varepsilon \cdot Y)$ for some $\varepsilon > 0$. Then*

$$\frac{\partial}{\partial \varepsilon}Q(q, \varepsilon) = \mathbb{E}\left[Y | X + \varepsilon \cdot Y = Q(q, \varepsilon)\right].$$

Proof. Denote by $f(x,y)$ the joint pdf of the pair (X,Y). We have

$$\mathbb{P}[X + \varepsilon \cdot Y > Q(q, \varepsilon)] = q \quad \Longleftrightarrow \quad \int_{-\infty}^{+\infty}\left(\int_{Q(q,\varepsilon)-\varepsilon y}^{+\infty} f(x,y)dx\right)dy = q.$$

Differentiation with respect to ε provides

$$\int_{-\infty}^{+\infty}\left(\frac{\partial Q(q,\varepsilon)}{\partial \varepsilon} - y\right)f(Q(q,\varepsilon) - \varepsilon y, y)dy = 0.$$

This leads to

$$\frac{\partial Q(q,\varepsilon)}{\partial \varepsilon} = \frac{\int_{-\infty}^{+\infty} yf(Q(q,\varepsilon) - \varepsilon y, y)dy}{\int_{-\infty}^{+\infty} f(Q(q,\varepsilon) - \varepsilon y, y)dy} = \mathbb{E}\left[Y | X + \varepsilon \cdot Y = Q(q, \varepsilon)\right]$$

which proves the assertion. □

Using this result we can prove that the first derivative

$$\frac{\partial}{\partial \varepsilon} \alpha_q (\mathbb{E}[L_N|X] + \varepsilon \cdot (L_N - \mathbb{E}[L_N|X]))|_{\varepsilon=0} = 0$$

in the Taylor expansion of the quantile vanishes since the idiosyncratic component conditional on the systematic component $\mathbb{E}[L_N|X]$ vanishes. Therefore, we set

$$X = \mathbb{E}[L_N|X] \quad \text{and} \quad Y = L_N - \mathbb{E}[L_N|X].$$

Then the above Lemma states

$$\begin{aligned}
\frac{\partial}{\partial \varepsilon} \alpha_q(X + \varepsilon \cdot Y)|_{\varepsilon=0} &= \mathbb{E}\left[Y | X + \varepsilon \cdot Y = \alpha_q(X + \varepsilon \cdot Y)\right]|_{\varepsilon=0} \\
&= \mathbb{E}\left[L_N - \mathbb{E}[L_N|X] \mid \mathbb{E}[L_N|X] = \alpha_q(X)\right] \\
&= \mathbb{E}[L_N] - \mathbb{E}[\mathbb{E}[L_N|X]] \\
&= 0
\end{aligned}$$

where we used the independence of $L_N - \mathbb{E}[L_N|X]$ and $\mathbb{E}[L_N|X]$. Finally, [76] proved that the remaining second derivative in the Taylor expansion, and thus the granularity adjustment, can be expressed as

$$GA_N = \frac{-1}{2h(\alpha_q(X))} \cdot \frac{d}{dx}\left(\frac{\sigma^2(x)h(x)}{\mu'(x)}\right)\bigg|_{x=\alpha_q(X)}, \quad (9.5)$$

where h is the density function of the systematic risk factor X.

This general framework can accommodate any definition of "loss". That is, we can measure the U_n on a mark-to-market basis or an actuarial basis, and either inclusive or exclusive of expected loss. The latter point is important in light of the separation of "total capital" (the concept used in CP2) into its EL and UL components in the final Basel II document. Say we measure the U_n and L_N inclusive of expected loss, but wish to define capital on a UL-only basis. Regulatory capital is measured as the qth quantile $\alpha_q(\mathbb{E}[L_N|X])$ of the expectation of the portfolio loss variable L_N conditional on the risk factor X (compare Chapter 4) under the assumption of an infinitely granular portfolio, i.e. when $L_N \to \mathbb{E}[L_N|X]$ as $N \to \infty$.[2] Let UL_N denote the "true" UL for the portfolio and let UL_N^{asympt} be the asymptotic approximation of the portfolio UL which assumes that the idiosyncratic risk is diversified away. Hence, when we split regulatory capital in its UL and EL components, we have

$$\alpha_q(\mathbb{E}[L_N|X]) = \text{UL}_N^{asympt} + \text{EL}_N^{asympt},$$

where $\text{EL}_N^{asympt} = \mathbb{E}[\mathbb{E}[L_N|X]]$. Then we obtain

---
[2] See Theorem 4.1.6.

$$GA_N = \alpha_q(L_N) - \alpha_q(\mathbb{E}[L_N|X]) = (\mathrm{UL}_N + \mathrm{EL}_N) - \left(\mathrm{UL}_N^{asympt} + \mathbb{E}[\mathbb{E}[L_N|X]]\right)$$
$$= \mathrm{UL}_N - \mathrm{UL}_N^{asympt}$$

because the unconditional expected loss ($\mathrm{EL}_N = \mathbb{E}[L_N]$) is equal to the expectation of the conditional expected loss ($\mathbb{E}[\mathbb{E}[L_N|X]]$). Put more simply, expected loss "washes out" of the granularity adjustment.

9.1.3 The Granularity Adjustment in a Single Factor CreditRisk+ Setting

The GA of [75] presents a first-order asymptotic approximation for the effect of diversification in large portfolios within the CreditRisk+ model of portfolio credit risk which we will now briefly explain.

In the GA formula, the expressions for $\mu(x)$, $\sigma^2(x)$ and $h(x)$ depend on the chosen model. For application of the GA in a supervisory setting, it would be desirable to base the GA on the same model as that which underpins the IRB capital formula. Unfortunately, this is not feasible for two reasons. First, the IRB formula is derived within a single-factor mark-to-market Vasicek model closest in spirit to KMV Portfolio Manager. The expressions for $\mu(x)$ and $\sigma^2(x)$ in such a model would be formidably complex. The effect of granularity on capital is sensitive to maturity, so simplification of the model to its default-mode counterpart (closest in spirit to a two-state CreditMetrics model) would entail a substantive loss of fidelity. Furthermore, even with that simplification, the resulting expressions for $\mu(x)$ and $\sigma^2(x)$ remain somewhat more complex than desirable for supervisory application. The second barrier to using this model is that the IRB formula is not fit to the model directly, but rather is linearized with respect to maturity. The "true" term-structure of capital charges in mark-to-market models tends to be strongly concave, so this linearization is not at all a minor adjustment. It is not at all clear how one would alter $\mu(x)$ and $\sigma^2(x)$ to make the GA consistent with the linearized IRB formula.

As fidelity to the IRB model cannot be imposed in a direct manner, [75] adopt an indirect strategy. They base the GA on a model chosen for the tractability of the resulting expressions, and then reparameterize the inputs in a way that restores consistency as much as possible. Their chosen model is an extended version of the single factor CreditRisk+ model that allows for idiosyncratic recovery risk. The CreditRisk+ model is a widely-used industry model for portfolio credit risk that was proposed by Credit Suisse Financial Products in 1997. For a general presentation of this model we refer to Chapter 6. As the CreditRisk+ model is an actuarial model of loss, one defines the loss rate as $U_n = \mathrm{LGD}_n \cdot D_n$. The systematic factor X generates correlation across obligor defaults by shifting the default probabilities. Conditional on $X = x$, the probability of default is[3]

---

[3] Compare equation (6.4).

$$\mathrm{PD}_n(x) = \mathrm{PD}_n \cdot (1 - w_n + w_n \cdot x),$$

where PD_n is the unconditional probability of default. The factor loading w_n controls the sensitivity of obligor n to the systematic risk factor. [75] assume that X is Gamma-distributed with mean 1 and variance $1/\xi$ for some positive ξ.[4] Finally, to obtain an analytical solution to the model, in CreditRisk$^+$ one approximates the distribution of the default indicator variable as a Poisson distribution (see Chapter 6 for details).

In the standard version of CreditRisk$^+$, the recovery rate is assumed to be known with certainty. The extended model allows LGD_n to be a random variable with expected value ELGD_n and variance VLGD_n^2. The LGD uncertainty is assumed to be entirely idiosyncratic, and therefore independent of X. For this model one can compute the $\mu(x)$ and $\sigma^2(x)$ functions. Therefore one defines at the instrument level the functions $\mu_n(x) = \mathbb{E}[U_n|x]$ and $\sigma_n^2(x) = \mathbb{V}[U_n|x]$. Due to the assumption of conditional independence, we have[5]

$$\mu(x) = \mathbb{E}[L_N|x] = \sum_{n=1}^{N} s_n \mu_n(x),$$

$$\sigma^2(x) = \mathbb{V}[L_N|x] = \sum_{n=1}^{N} s_n^2 \sigma_n^2(x).$$

In CreditRisk$^+$, the $\mu_n(x)$ function is simply

$$\mu_n(x) = \mathrm{ELGD}_n \cdot \mathrm{PD}_n(x) = \mathrm{ELGD}_n \cdot \mathrm{PD}_n \cdot (1 - w_n + w_n \cdot x).$$

For the conditional variance, we have

$$\begin{aligned}\sigma_n^2(x) &= \mathbb{E}[\mathrm{LGD}_n^2 \cdot D_n^2 | x] - \mathrm{ELGD}_n^2 \cdot \mathrm{PD}_n(x)^2 \\ &= \mathbb{E}[\mathrm{LGD}_n^2] \cdot \mathbb{E}[D_n^2|x] - \mu_n(x)^2.\end{aligned} \quad (9.6)$$

As D_n given X is assumed to be Poisson distributed, we have $\mathbb{E}[D_n|X] = \mathbb{V}[D_n|X] = \mathrm{PD}_n(X)$, which implies

$$\mathbb{E}[D_n^2|X] = \mathrm{PD}_n(X) + \mathrm{PD}_n(X)^2.$$

For the term $\mathbb{E}[\mathrm{LGD}_n^2]$ in the conditional variance, we can substitute

$$\mathbb{E}[\mathrm{LGD}_n^2] = \mathbb{V}[\mathrm{LGD}_n] + \mathbb{E}[\mathrm{LGD}_n]^2 = \mathrm{VLGD}_n^2 + \mathrm{ELGD}_n^2.$$

This leads us to

$$\begin{aligned}\sigma_n^2(x) &= \left(\mathrm{VLGD}_n^2 + \mathrm{ELGD}_n^2\right) \cdot \left(\mathrm{PD}_n(X) + \mathrm{PD}_n(X)^2\right) - \mu_n(x)^2 \\ &= C_n \mu_n(x_q) + \mu_n(x_q)^2 \cdot \frac{\mathrm{VLGD}_n^2}{\mathrm{ELGD}_n^2}\end{aligned} \quad (9.7)$$

---
[4] Note that we must have $\mathbb{E}[X] = 1$ in order that $\mathbb{E}[\mathrm{PD}_n(X)] = \mathrm{PD}_n$.
[5] Recall here the equation (9.1) for the portfolio loss L_N.

9.1 A Granularity Adjustment for the ASRF Model

where $\mathcal{C}_n$ is defined as

$$\mathcal{C}_n = \frac{\mathrm{ELGD}_n^2 + \mathrm{VLGD}_n^2}{\mathrm{ELGD}_n}. \tag{9.8}$$

We substitute the Gamma pdf $h(x)$ and the expressions for $\mu(x)$ and $\sigma^2(x)$ into equation (9.5), and then evaluate the derivative in that equation at $x = \alpha_q(X)$. The resulting formula depends on the instrument-level parameters PD_n, w_n, ELGD_n and VLGD_n.

[75] then reparameterize the inputs in terms of EL and UL capital requirements. Let $\mathcal{R}_n$ be the EL reserve requirement as a share of EAD for instrument n. In the default-mode setting of CreditRisk$^+$, this is simply

$$\mathcal{R}_n = \mathrm{ELGD}_n \cdot \mathrm{PD}_n.$$

Let $\mathcal{K}_n$ be the UL capital requirement as a share of EAD. In CreditRisk$^+$, this is

$$\mathcal{K}_n = \mathbb{E}[U_n | X = \alpha_q(X)] = \mathrm{ELGD}_n \cdot \mathrm{PD}_n \cdot w_n \cdot (\alpha_q(X) - 1) \tag{9.9}$$

When we substitute $\mathcal{R}_n$ and $\mathcal{K}_n$ into the CreditRisk$^+$ GA, we find that the PD_n and w_n inputs can be eliminated. We arrive at the formula

$$GA_N = \frac{1}{2\mathcal{K}^*} \sum_{n=1}^{N} s_n^2 \left[\left(\delta \mathcal{C}_n (\mathcal{K}_n + \mathcal{R}_n) + \delta (\mathcal{K}_n + \mathcal{R}_n)^2 \cdot \frac{\mathrm{VLGD}_n^2}{\mathrm{ELGD}_n^2} \right) \right. \\ \left. - \mathcal{K}_n \left(\mathcal{C}_n + 2(\mathcal{K}_n + \mathcal{R}_n) \cdot \frac{\mathrm{VLGD}_n^2}{\mathrm{ELGD}_n^2} \right) \right], \tag{9.10}$$

where $\mathcal{K}^* = \sum_{n=1}^{N} s_n \mathcal{K}_n$ is the required capital per unit exposure for the portfolio as a whole and where

$$\delta \equiv (\alpha_q(X) - 1) \cdot \left(\xi + \frac{1-\xi}{\alpha_q(X)} \right).$$

Note that the expression for δ depends only on model parameters, not on data inputs, so δ is a regulatory parameter. It is through δ that the variance parameter ξ influences the GA. In the CP2 version, we set $\xi = 0.25$. Assuming that the target solvency probability is $q = 0.999$, this setting implies $\delta = 4.83$. For policy purposes, it is worthwhile to note that setting $\xi = 0.31$ would be well within any reasonable bounds on this parameter, and would yield the parsimonious integer value $\delta = 5$.

The volatility of LGD (VLGD) neither is an input to the IRB formula, nor is it restricted in any way within the IRB model. Therefore, we need to impose a regulatory assumption on VLGD in order to avoid burdening banks with additional data requirements. [75] impose the relationship as found in the CP2 version of the GA

$$\text{VLGD}_n^2 = \gamma \, \text{ELGD}_n(1 - \text{ELGD}_n), \tag{9.11}$$

where the regulatory parameter γ is between 0 and 1. When this specification is used in industry models such as CreditMetrics and KMV Portfolio Manager, a typical setting is $\gamma = 0.25$.

The GA formula can be simplified somewhat. The quantities $\mathcal{R}_n$ and $\mathcal{K}_n$ are typically small, so terms that are products of these quantities can be expected to contribute little to the GA. If these second-order terms are dropped, [75] arrive at the simplified formula

$$\widetilde{GA}_N = \frac{1}{2\mathcal{K}^*} \sum_{n=1}^{N} s_n^2 \mathcal{C}_n \left(\delta(\mathcal{K}_n + \mathcal{R}_n) - \mathcal{K}_n \right). \tag{9.12}$$

Here and henceforth, we use the tilde to indicate this simplified GA formula. The accuracy of this approximation to the "full" GA is discussed in [75].

9.1.4 Data on German Bank Portfolios

To show the impact of the granularity adjustment on economic capital, [75] apply the GA to realistic bank portfolios. They use data from the *German credit register,* which includes all bank loans greater or equal to 1.5 Million Euro. This data set has been matched to the firms' balance sheet data to obtain obligor specific PDs. The resulting portfolios are much smaller than the portfolios reported in the German credit register, however, there are still a number of banks with more than 300 exposures in this matched data set which they consider as an appropriate size for calculating the GA. The banks are grouped in large, medium, small and very small banks where large refers to a bank with more than 4000 exposures, medium refers to one with 1000 – 4000 exposures, small refers to a bank with 600 – 1000 exposures and very small to a bank with 300 – 600 exposures.

To accommodate privacy restrictions on these data, [75] aggregate portfolios for three different banks into a single data set. They then sort the loans by exposure size and remove every third exposure. The resulting portfolio of 5289 obligors is still realistic in terms of exposure and PD distribution and is similar in size to some of the larger portfolios in the matched data set of the German credit register and the firm's balance sheet data. The mean of the loan size distribution is 3973 thousand Euros and the standard deviation is 9435 thousand Euros. Quantiles of the loan size distribution are reported in Table 9.1. Henceforth, we refer to this portfolio as "Portfolio A."

Figure 9.2 shows the PD distribution for the aggregated portfolio A for different PD categories which we denote here by S&P's common rating grades. The PD ranges for the different rating grades are listed in Table 9.2 below.

The average PD of the data set is 0.43% and hence lower than the average PD of a portfolio of a smaller or medium sized bank in Germany, which is approximately 0.8% [86, p. 8]. Moody's, for example, underlies average net

9.1 A Granularity Adjustment for the ASRF Model   85

Table 9.1. Exposure distribution in Portfolio A (in thousand Euros)

Level Quantile	
5%	50.92
25%	828.80
50%	1811.75
75%	3705.50
95%	13637.36

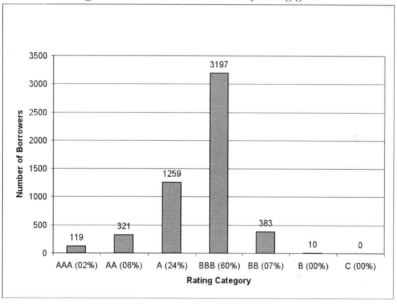

Fig. 9.2. Borrower distribution by rating grade

Table 9.2. PD ranges associated with rating buckets

Rating Grade	PD Ranges in %
AAA	$PD \leq 0.02$
AA	$0.02 \leq PD \leq 0.06$
A	$0.06 \leq PD \leq 0.18$
BBB	$0.18 \leq PD \leq 1.06$
BB	$1.06 \leq PD \leq 4.94$
B	$4.94 \leq PD \leq 19.14$
C	$19.14 \leq PD$

loan provisions of 0.77% for German banks during the period 1989 – 1999 (see [86, p. 7]), which is more than two times the average loss of the firms in our sample during the same period. Approximately 70% of the portfolio in our data set belongs to the investment grade domain (i.e., rated BBB or better) and the remaining 30% to the subinvestment grade. In smaller or medium sized banks in Germany the percentage of investment grade exposures in a portfolio is approximately 37% (compare [118, p. 2]). As a consequence the value of the GA in the aggregated portfolio A will be smaller than the GA in a true bank portfolio of similar exposure distribution.

The data set does not contain information on LGD, so [75] impose the Foundation IRB assumption of ELGD = 0.45.

9.1.5 Numerical Results

In Table 9.3, we present granularity adjustments calculated on real bank portfolios varying in size and degree of heterogeneity. As one would expect, the GA is invariably small (12 to 14 basis points) for the largest portfolios, but can be substantial (up to 161 basis points) for the smallest. The table demonstrates the strong correlation between Herfindahl-Hirschman index and GA across these portfolios, though of course the correspondence is not exact as the GA is sensitive to credit quality as well. As a reference portfolio, we included a portfolio with 6000 loans each of PD = 0.01 and ELGD = 0.45 and of homogeneous EAD. The GA for the largest real portfolio is roughly six times as large as the GA for the homogeneous reference portfolio, which demonstrates the importance of portfolio heterogeneity in credit concentrations.

Table 9.3. Granularity Adjustment for real bank portfolios

Portfolio	Number of Exposures	HHI	GA (in %)
Reference	6000	0.00017	0.018
Large	> 4000	< 0.001	0.12 − 0.14
Medium	1000 − 4000	0.001 − 0.004	0.14 − 0.36
Small	600 − 1000	0.004 − 0.011	0.37 − 1.17
Very Small	300 − 600	0.005 − 0.015	0.49 − 1.61

[75] also computed the VaR in the CreditRisk$^+$ model and the relative add-on for the GA on the VaR. For a large portfolio this add-on is 3% to 4% of VaR. For a medium sized bank the add-on lies between 5% and 8% of VaR. In a study based on applying a default-mode multi-factor CreditMetrics model to US portfolio data, [25] find that name concentration accounts for between 1% and 8% of VaR depending on the portfolio size. These results are quite close to the ones of [75] for the GA, despite the difference in model and data.

9.1 A Granularity Adjustment for the ASRF Model

Table 9.4. GA as percentage add-on to RWA

Portfolio	Number of Exposures	Relative Add-On for RWA (in %)
Reference	6000	0.003
Large	> 4000	0.04
Medium	1000 − 4000	0.04 − 0.10
Small	300 − 1000	0.17 − 0.32

Table 9.4 shows the relative add-on for the granularity adjustment on the Risk Weighted Assets (RWA) of Basel II for small, medium and large portfolios as well as for the reference portfolio with 6000 exposures of size one. The reference portfolio is used to point out the influence of the GA even for large portfolios which one would intuitively consider as highly granular. For the reference portfolio of 6000 exposures of size one with homogeneous PD = 1% and ELGD = 45% the GA is approximately 0.018% and the IRB capital charge is 5.86%. Thus the add-on due to granularity is approximately 0.3% and the economic capital to capture both systematic risk and risk from single-name concentration is 5.878% of the total portfolio exposure. For the real bank portfolios of our data set the add-on for the GA is higher than for the reference portfolio although it is still quite small for large and even for some of the medium sized bank portfolios. For smaller portfolios with 300 to 1000 exposures the add-on for the GA is more significant.

Figure 9.3 shows the dependence of the simplified GA on the default probability. Each point of the curve represents a homogeneous portfolio of $N = 1000$ borrowers of the given PD. Dependence on portfolio quality is non-negligible, particularly for lower-quality portfolios. Such dependence cannot be accomodated naturally and accurately in ad-hoc methods of granularity adjustment based on exposure HHI.

The sensitivity of the GA to the variance parameter ξ of the systematic factor X is explored in Figure 9.4. We see that the granularity adjustment is strictly increasing in ξ, and that the degree of sensitivity is not negligible. Increasing ξ from 0.2 to 0.3 causes a 10% increase in the GA for Portfolio A. Parameter uncertainty of this sort is a common predicament in modeling economic capital, and here we might recall the challenges faced by the Basel Committee in setting a value for the asset correlation parameter (ϱ). While the absolute magnitude of the GA is sensitive to ξ, its relative magnitude across bank portfolios is much less so. In this sense, the proper functioning of the GA as a supervisory tool does not materially depend on the precision with which ξ is calibrated.

[75] also verified the accuracy of the simplified granularity adjustment $\widetilde{\text{GA}}$ as an approximation to the "full" GA of equation (9.10). They constructed

Fig. 9.3. Effect of credit quality on simplified GA

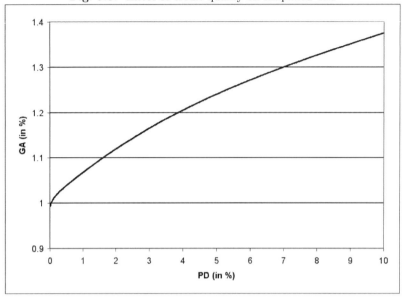

six stylized portfolios of different degrees of exposure concentrations. Each portfolio consists of $N = 1000$ exposures and has constant PD and ELGD fixed at 45%. Portfolio P0 is completely homogeneous whereas portfolio P50 is highly concentrated since the largest exposure $A_{1000} = 1000^{50}$ accounts for 5% of the total exposure of the portfolio. The values for both the simplified $\widetilde{GA}_N$ and the full GA_N for each of these portfolios are listed in Table 9.5. We see that the approximation error increases with concentration and with PD. For realistic portfolios, the error is trivial. Even for the case of portfolio P10 and PD = 4%, the error is only 3 basis points. The error grows to 12 basis points in the extreme example of P50 and PD = 4%, but even this remains small relative to the size of the GA.

9.1.6 Summary

The paper of [75] sets forth a granularity adjustment for portfolio credit VaR that accounts for a risk that is not captured by the Pillar 1 capital requirements of the Basel II IRB approach. The discussed GA is a revision and an extension of the methodology first introduced in the Basel II Second Consultative Paper (CP2) [12]. The revision incorporates some technical advances as well as modifications to the Basel II rules since CP2.

[75] also introduce an "upper bound" methodology that addresses the most significant source of operational burden associated with the assessment of

9.1 A Granularity Adjustment for the ASRF Model

Fig. 9.4. Effect of the variance of the systematic factor on simplified GA

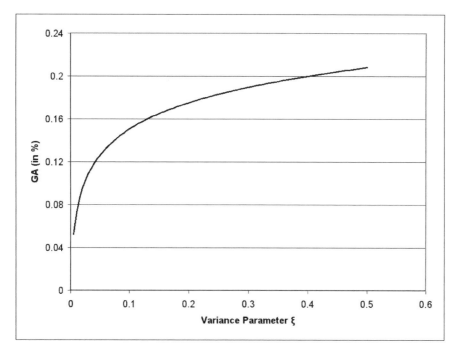

Table 9.5. Approximation error of the simplified $\widetilde{GA}_N$

Portfolio	P0	P1	P2	P10	P50
PD = 1%					
Exposure A_n	1	n	n^2	n^{10}	n^{50}
$\widetilde{GA}$ in %	0.107	0.142	0.192	0.615	2.749
GA in %	0.109	0.146	0.197	0.630	2.814
PD = 4%					
Exposure A_n	1	n	n^2	n^{10}	n^{50}
$\widetilde{GA}$ in %	0.121	0.161	0.217	0.694	3.102
GA in %	0.126	0.168	0.227	0.726	3.243

residual idiosyncratic risk in the portfolio (whether through the proposed GA or by any other rigorous methodology). For many banks, this approach would permit dramatic reductions in data requirements at modest cost in additional capital requirements. We did not discuss the upper bound methodology in this section since we only wish to present and compare different methods for measuring name concentration in these lecture notes. For more details and results that are particularly interesting for practical application we refer to the original paper at this point.

A detailed discussion of the GA approach reviewed in this section and a comparison study of the different methods for measuring name concentration is presented at the end of this chapter in Section 9.4.

9.2 The Semi-Asymptotic Approach

As already mentioned in the beginning of this chapter, [52] offer two approaches to measure name concentration; a granularity adjustment based on a one-factor default-mode CreditMetrics model and a semi-asymptotic approach.

Their granularity adjustment approach is based on the same idea as the GA we presented in the previous section. However, it is formulated within a different credit portfolio model. We will briefly discuss the advantages and disadvantages of this approach in the comparison study at the end of this chapter.

In this section we will give a short overview of the semi-asymptotic approach which relies on a limiting procedure that is applied only to a part of the portfolio.

9.2.1 The General Framework

For our portfolio of N risky loans introduced in Section 1.2 we fix a stochastic model to describe the portfolio loss variable L_N. The semi-asymptotic approach of [52] is formulated within a one-factor default-mode CreditMetrics model such that the portfolio loss L_N is given by

$$L_N = \sum_{n=1}^{N} s_n \mathbf{1}_{\{\sqrt{\varrho_n} X + \sqrt{1-\varrho_n}\varepsilon_n \leq c_n\}},$$

where the asset correlation $0 < \varrho_n < 1$, $n = 1, \ldots, N$, measures the degree of sensitivity of obligor n with respect to the systematic risk factor X. The default thresholds $c_n \geq 0$ are constants for $n = 1, \ldots, N$. The systematic risk factor X as well as the idiosyncratic risk factors $\varepsilon_1, \ldots, \varepsilon_N$ are independent random variables with continuous distributions.

In the semi-asymptotic approach, [52] consider the special case where $\varrho_1 = \tau$, $c_1 = a$ but $\varrho_n = \varrho$ and $c_n = c$ for $n > 1$. The exposure shares s_n in general depend on the total portfolio exposure and, therefore, also on the number N of exposures in the portfolio. In [52], however, it is assumed that the exposure share of obligor 1 is constant $s_1 = s$ for all N. The remaining exposure shares $s_2, s_3, \ldots$ are supposed to fulfill

$$\lim_{N \to \infty} \sum_{n=2}^{N} s_n^2 = 0. \tag{9.13}$$

9.2 The Semi-Asymptotic Approach

Hence, our model for the portfolio loss L_N reduces to

$$L_N = s \cdot \mathbf{1}_{\{\sqrt{\tau}X + \sqrt{1-\tau}\varepsilon \leq a\}} + (1-s) \sum_{n=2}^{N} s_n \cdot \mathbf{1}_{\{\sqrt{\varrho}X + \sqrt{1-\varrho}\varepsilon_n \leq c\}} \quad (9.14)$$

with $\sum_{n=2}^{N} s_n = 1$. Note that also $s + (1-s) \sum_{n=2}^{N} s_n = 1$. In this situation all assets in the portfolio are influenced by the single systematic factor X to the extend ϱ, only asset s is influenced by the systematic factor X to a different degree, namely τ. However, a natural choice for τ might also be $\tau = \varrho$, the mean portfolio asset correlation. For large N the portfolio becomes more and more fine-grained. For N tending to ∞, this model setting leads to the *semi-asymptotic* percentage loss function

$$L(s) = s \cdot D + (1-s) \cdot Y \quad (9.15)$$

with default indicator variable

$$D = \mathbf{1}_{\{\sqrt{\tau}X + \sqrt{1-\tau}\varepsilon \leq a\}} \quad (9.16)$$

and random variable

$$Y = \mathbb{P}\left[\varepsilon \leq \frac{c - \sqrt{\varrho}X}{\sqrt{1-\varrho}}\right]. \quad (9.17)$$

Here we used the same reasoning as in Section 4.3, equation (4.7), where we saw that, in the limit $N \to \infty$, the percentage portfolio loss L_N tends to the conditional default probability (4.6). The transition from equation (9.14) to (9.15) is called *semi-asymptotic* since the first term is constant for $N \to \infty$ and only the second term changes with N.

Definition 9.2.1 (Semi-Asymptotic Capital Charge)
The quantity

$$s \cdot \mathbb{P}[D = 1 | L(s) = \alpha_q(L(s))]$$

is called the *semi-asymptotic capital charge* (at level q) of the loan with exposure share s (as percentage of total portfolio exposure) and default indicator D as in (9.16).[6]

These capital charges are not portfolio invariant, however, they can take into account concentration effects. Moreover, their dependence on the exposure share s is not linear since the factor $\mathbb{P}[D = 1 | L(s) = \alpha_q(L(s))]$ depends on s as well. Note also that (9.15) can be understood as a two-asset portfolio model in which case Definition 9.2.1 is the standard definition of VaR contributions (compare also [94] and [119]).

Now assume that the conditional distribution functions F_0 and F_1 of Y given $D = 0$ and $D = 1$, respectively, have densities that are continuous and concentrated on the interval $(0,1)$ and call their densities f_0 and f_1. Then

---
[6] See [52], Definition 3.1.

$$\mathbb{P}\left[Y \leq y | D = i\right] = F_i(y) = \int_{-\infty}^{y} f_i(t) dt, \quad \text{for } y \in \mathbb{R}, \, i \in \{0,1\}.$$

In this case, [52] computed the distribution function and the density of $L(s)$ as

$$\mathbb{P}\left[L(s) \leq z\right] = \text{PD} \cdot F_1\left(\frac{z-s}{1-s}\right) + (1 - \text{PD}) \cdot F_0\left(\frac{z}{1-s}\right)$$

and

$$\frac{\partial \mathbb{P}\left[L(s) \leq z\right]}{\partial z} = (1-s)^{-1}\left(\text{PD} \cdot f_1\left(\frac{z-s}{1-s}\right) + (1 - \text{PD}) \cdot f_0\left(\frac{z}{1-s}\right)\right),$$

respectively, where $\text{PD} = \mathbb{P}[D = 1]$ denotes the default probability of the loan under consideration. These formulas provide a method to compute the quantile $\alpha_q(L(s))$ numerically. Moreover, the conditional probability in Definition 9.2.1 can be computed as

$$\mathbb{P}\left[D = 1 | L(s) = z\right] = \frac{\text{PD} \cdot f_1\left(\frac{z-s}{1-s}\right)}{\text{PD} \cdot f_1\left(\frac{z-s}{1-s}\right) + (1 - \text{PD}) \cdot f_0\left(\frac{z}{1-s}\right)}.$$

The following proposition states the relation between the default probability PD and the quantile level q.[7]

Proposition 9.2.2 *For $L(s)$ defined by (9.15) with $F_1(0) = 0$, we have*

$$\lim_{s \to 1} \alpha_q(L(s)) = \begin{cases} 1, & \text{for } \text{PD} > 1 - q \\ 0, & \text{for } \text{PD} \leq 1 - q \end{cases}$$

For the proof see [52], Proposition 3.2. They also derive a limiting result for the behavior of the quantile $\alpha_q(L(s))$ under slightly stronger assumptions.[8]

Proposition 9.2.3 *Assume that the densities f_0 and f_1 are positive in $(0,1)$. Then for $L(s)$ defined by (9.15) we have*

$$\lim_{s \to 1} \mathbb{P}\left[D = 1 | L(s) = \alpha_q(L(s))\right] = \begin{cases} 1, & \text{for } \text{PD} > 1 - q, \\ 0, & \text{for } \text{PD} < 1 - q. \end{cases}$$

For the proof see [52], Proposition 3.3. This result shows that putting all risk into an event with very small probability can quite drastically reduce capital charges which is a general problem of VaR based measures. Therefore, [52] suggest to use another risk measure as, for example, Expected Shortfall, in which case the conditional probability $\mathbb{P}\left[D = 1 | L(s) = \alpha_q(L(s))\right]$ would have to be replaced by $\mathbb{P}\left[D = 1 | L(s) \geq \alpha_q(L(s))\right]$.

---

[7] See [52], Proposition 3.2.
[8] See [52], Proposition 3.3.

9.2.2 Numerical Results

As an example, [52] assume that the portfolio modeled by Y has a moderate credit standing expressed by its expected loss $\mathbb{E}[Y] = 0.025 = \Phi(c)$. They impose quite a strong exposure to systematic risk by choosing the asset correlation $\varrho = 0.1$ and, for simplicity they set $\tau = \varrho$. The additional loan s has a very low probability of default $\text{PD} = \mathbb{P}[D = 1] = 0.002 = \Phi(a)$.

[52] explore the dependence of the portfolio loss variable $L(s)$ on the relative weight s of the new loan by comparing the true VaR at level 99.9% (computed numerically) to the GA version of their paper and to the Basel II risk weight function. Their results show that, up to a level of 5% relative weight of the new loan, both the GA approach as well as the Basel II approach yield very precise approximations to the true portfolio VaR. In this example, [52] show that the granularity adjustment approach can yield misleading results if there is a non-negligible concentration in the portfolio. Their alternative approach, the semi-asymptotic approach, however, can also come up with counter-intuitive results if the default probability on the loan under consideration is very small. To avoid this shortcoming they suggest to use a different risk measure as, for example, expected shortfall.

9.3 Methods Based on the Saddle-Point Approximation

In credit risk management one is particularly interested in the portfolio loss distribution. As the portfolio loss is usually modeled as the sum of random variables, the main task is to evaluate the probability density function (pdf) of such a sum of random variables. When these random variables are independent, then the pdf is just the convolution of the pdfs of the individual obligor loss distributions. The evaluation of this convolution, however, is computationally intensive, on the one hand, and, on the other hand, analytically in most cases quite complicated since the loss distribution usually does not possess a closed form solution. However, in some cases, moments can be computed which allows to approximate the loss distribution based on the moment generating function. In such situations a technique, which is frequently used, is the *Edgeworth expansion* method which works quite well in the center of a distribution. In the tails, however, this technique performs poorly. The *saddle-point expansion* can be regarded as an improvement of the Edgeworth expansion. It is a technique for approximating integrals used in physical sciences, engineering and statistics. [99] were the first to apply the saddle-point approximation method to credit risk and in particular to the computation of credit VaR. They derived a formula for calculating the marginal impact of a new exposure in a given portfolio. Therefore, the method can be used to quantify the concentration risk of a portfolio. The method is based on the moment generating function (MGF) of the portfolio loss distribution. The stationary points of the MGF (which can be regarded as saddle points) contain lots of information about the shape of the loss distribution. The saddle-points of the MGF are quite easy to compute, and can be used to derive the shape of the tail of the loss distribution without using time-consuming Monte Carlo simulation. Moreover, it can be shown that the saddle-point approximation method is accurate and robust in cases where standard methods perform poorly, and vice-versa.

In this section we first review the saddle-point method and then show an application of this approach to the measurement of name concentration risk. Here we follow mainly the papers [99] and [102]. In Section 9.4 we then compare this method with the granularity adjustment technique and show that both methods coincide in case of the single-factor CreditRisk$^+$ model.

9.3.1 The General Framework

Recall that the moment generating function (MGF) of a random variable Y with density f_Y is defined as the analytic function

$$M_Y(s) = \mathbb{E}\left[e^{sY}\right] = \int e^{st} f_Y(t) dt$$

of a complex variable s, provided that the integral exists.[9] Inverting the above formula gives[10]

$$f_Y(t) = \frac{1}{2\pi i} \int_{-i\infty}^{+i\infty} e^{-st} M_Y(s) ds \qquad (9.18)$$

where the path of integration is the imaginary axis. As already mentioned before, the MGF of a sum of independent random variables is given by the product of the MGFs of the individual random variables. The saddle-point method now provides a tool to derive the pdf of such a sum of random variables by approximating the integral in the inversion formula (9.18).

Define the *cumulant generating function* (CGF) of Y as $K_Y(s) = \ln M_Y(s)$. Thus we can rewrite equation (9.18) as

$$f_Y(t) = \frac{1}{2\pi i} \int_{-i\infty}^{+i\infty} \exp\left(K_Y(s) - st\right) ds. \qquad (9.19)$$

The *saddle-points* are the points at which the terms in the exponential are stationary, i.e. the points s for which

$$\frac{\partial}{\partial s}\left(K_Y(s) - st\right) = \frac{\partial K_Y(s)}{\partial s} - t = 0. \qquad (9.20)$$

---

[9] See Definition 6.3.2.
[10] As the MGF of a random variable Y is just the two sided Laplace transform of its density function f_Y, we can invert the MGF by inversion of the Laplace transform. See, for example, [37] for details on Laplace transforms.

9.3 Methods Based on the Saddle-Point Approximation

Having found these points the saddle-point approximation method consists in applying a second-order Taylor expansion to the term $K_Y(s) - st$ and then computing the resulting Gaussian integral as will be demonstrated in the following.[11] By equation (9.20) for fixed t the saddle-point $\hat{t}$ satisfies

$$\frac{\partial K_Y(s)}{\partial s}\bigg|_{s=\hat{t}} = \frac{1}{M_Y(\hat{t})} \cdot \frac{\partial M_Y(s)}{\partial s}\bigg|_{s=\hat{t}} = t. \qquad (9.21)$$

It can be shown that for each t in the distribution of Y, there exists a unique $\hat{t}$ on the real axis that solves (9.21) since the cumulant generating function is convex[12]. Note that for $\hat{t} = 0$ we have

$$\frac{\partial K_Y(s)}{\partial s}\bigg|_{s=0} = \frac{1}{M_Y(0)} \cdot \frac{\partial M_Y(s)}{\partial s}\bigg|_{s=0} = \int u \cdot f_Y(u)du = \mathbb{E}[Y],$$

implying that $\hat{t} = 0$ is the saddle-point corresponding to the value $t = \mathbb{E}[Y]$. If we think of t being a quantile of the distribution of Y then, due to equation (9.21), we can relate quantiles t of the distribution of Y to K_Y evaluated at the saddle-point $\hat{t}$. This is an important feature for applications in credit risk. Moreover, this is also one reason why we obtain analytical expressions for the portfolio VaR.

Applying a second-order Taylor expansion to the term $K_Y(s) - st$ around $\hat{t}$ in equation (9.19) and then computing the resulting Gaussian integral we obtain

$$\begin{aligned}
f_Y(t) &= \frac{1}{2\pi i} \int_{-i\infty}^{+i\infty} \exp\left(K_Y(s) - st\right) ds \\
&\approx \frac{\exp\left(K_Y(\hat{t}) - t\hat{t}\right)}{2\pi i} \int_{-i\infty}^{+i\infty} \exp\left(\frac{1}{2}(s-\hat{t})^2 K_Y''(\hat{t})\right) ds \qquad (9.22) \\
&= \frac{\exp\left(K_Y(\hat{t}) - t\hat{t}\right)}{\sqrt{2\pi K_Y''(\hat{t})}}.
\end{aligned}$$

Here we used that the first term in the Taylor expansion vanishes at $\hat{t}$ due to the definition of the saddle-point $\hat{t}$. Note that by applying equation (9.19) we can compute the tail probability as

---

[11] See e.g. [34] or [83] for further information.
[12] Compare [83].

$$\mathbb{P}(Y > t) = \int_t^\infty f_Y(u)\,du$$

$$= \int_t^\infty \left(\frac{1}{2\pi i} \int_{-i\infty}^{+i\infty} \exp\left(K_Y(s) - su\right) ds \right) du$$

$$= \frac{1}{2\pi i} \int_{-i\infty,(0+)}^{+i\infty} \left[\frac{\exp(K_Y(s) - su)}{-s} \right]_{u=t}^\infty ds$$

$$= \frac{1}{2\pi i} \int_{-i\infty,(0+)}^{+i\infty} \frac{\exp(K_Y(s) - st)}{s} ds$$

where the integration path is the imaginary axis and the notation (0+) denotes that the contour runs to the right of the origin to avoid the pole there. Thus by applying formula (9.22) the tail probability of Y can be recovered from the CGF by a contour integral of the form

$$\mathbb{P}(Y > t) = \frac{1}{2\pi i} \int_{-i\infty,(0+)}^{+i\infty} \frac{\exp(K_Y(s) - st)}{s} ds \qquad (9.23)$$

$$\approx \frac{\exp(K_Y(\hat{t}) - t\hat{t})}{2\pi i} \int_{-i\infty,(0+)}^{+i\infty} \frac{\exp\left(\frac{1}{2}(s - \hat{t})^2 K_Y''(\hat{t})\right)}{s} ds.$$

This can be computed further and we finally obtain[13]

$$\mathbb{P}(Y > t) \approx \begin{cases} \exp\left(K_Y(\hat{t}) - t\hat{t} + \frac{1}{2}\hat{t}^2 K_Y''(\hat{t})\right) \Phi\left(-\sqrt{\hat{t}^2 K_Y''(\hat{t})}\right) & \text{for } t > \mathbb{E}[Y] \\ \frac{1}{2} & \text{for } t = \mathbb{E}[Y] \\ 1 - \exp\left(K_Y(\hat{t}) - t\hat{t} + \frac{1}{2}\hat{t}^2 K_Y''(\hat{t})\right) \Phi\left(-\sqrt{\hat{t}^2 K_Y''(\hat{t})}\right) & \text{for } t < \mathbb{E}[Y] \end{cases}$$
(9.24)

where Φ denotes the cumulative normal distribution function.

9.3.2 Application to Name Concentration Risk

In this sequel we apply the saddle-point approximation method to measure name concentration risk within a one-factor framework where the single systematic risk factor is denoted by X. Therefore, consider a portfolio of M risky loans indexed by $m = 1, \ldots, M$.[14] Denote by s_m the exposure share of obligor m and by $\mu_m(x)$ and $\sigma_m^2(x)$ the conditional mean and variance, respectively,

---
[13] See [99] for details.
[14] We deviate here from our usual notation as the variable N, which described the number of loans in our former portfolios, will be used later on in the construction of new portfolios which become more and more fine-grained as N increases.

9.3 Methods Based on the Saddle-Point Approximation

of the loss ratio of obligor m given $X = x$. Then we obtain for the expectation $\mu(x)$ and variance $\sigma^2(x)$ of the portfolio loss L_1 of this (first) portfolio conditional on the systematic risk factor X

$$\mu(X) = \mathbb{E}[L_1|X] = \sum_{m=1}^{M} s_m \mu_m(X)$$

$$\sigma^2(X) = \mathbb{V}[L_1|X] = \sum_{m=1}^{M} s_m^2 \sigma_m^2(X).$$

Starting with this portfolio of M risky loans, [102] construct a sequence of new portfolios such that the expected loss in the portfolios remains constant. In each step the exposures in the former portfolio are divided by the iteration step number, i.e. in step N there exist NM loans in the portfolio and each N of them are identical in size, namely s_m/N for $m = 1, \ldots, M$. Let L_N denote the loss distribution of the Nth portfolio constructed in this way. Note that each of these portfolios satisfies Assumptions 4.1.2 and 4.1.3. Theorem 4.1.4 then states that the portfolio loss variable L_N tends to the systematic loss distribution almost surely for N large, i.e.

$$L_N - \mathbb{E}[L_N|X] \longrightarrow 0 \quad \text{almost surely as} \quad N \to \infty.$$

Recall that the difference in terms of percentiles between L_N and the conditional expected loss $\mathbb{E}[L_N|X]$ is the granularity adjustment. In this one-factor model we have conditional on $X = x$

$$\mathbb{E}[L_N|X] = \mu(X) \quad \text{and}$$

$$\mathbb{V}[L_N|X] = N \cdot \sum_{m=1}^{M} \left(\frac{s_m}{N}\right)^2 \cdot \sigma_m^2(X) = \sigma^2(X)/N$$

for all $N \geq 1$. Thus the conditional mean stays constant for all portfolios while the conditional variance gets divided by the number N of the iteration step. For convenience we introduce u where $u = 1/N$. Let Z be a mean zero and unit variance random variable. Then we can express the loss distribution L_N of the Nth portfolio as

$$L_u = \mathbb{E}[L_N|X] + \sqrt{u} \cdot \sigma(X) \cdot Z. \tag{9.25}$$

We now want to derive a saddle-point equivalent to the GA formula based on the analytical approximation (9.23) to VaR.[15] If we replace the random variable Y in equation (9.23) by the loss variable L_u and then differentiate with respect to u we obtain

$$\frac{\partial}{\partial u}\mathbb{P}(L_u > t) = \frac{1}{2\pi i}\int_{-i\infty,(0+)}^{+i\infty} \left(\frac{1}{s}\frac{\partial K_{L_u}(s)}{\partial u} - \frac{\partial t}{\partial u}\right) \cdot \exp\left(K_{L_u}(s) - st\right) ds.$$

---

[15] Here we follow the derivations in [101], but modify their approach for comparison reasons with the GA methodology.

We want to study how the VaR $\alpha_q(L_u)$ for some fixed quantile level q depends on u. Thus we consider the case $t = \alpha_q(L_u)$. In this situtation the left hand side in the above equation vanishes since $\mathbb{P}(L_u > \alpha_q(L_u)) \equiv q$ and after some rearranging we obtain

$$\frac{\partial \alpha_q(L_u)}{\partial u} = \frac{\int_{-i\infty,(0+)}^{+i\infty} \frac{1}{s} \cdot \frac{\partial K_{L_u}(s)}{\partial u} \cdot \exp\left(K_{L_u}(s) - s\alpha_q(L_u)\right) ds}{\int_{-i\infty,(0+)}^{+i\infty} \exp\left(K_{L_u}(s) - s\alpha_q(L_u)\right) ds}.$$

Recall from Subsection (9.1.2) that the granularity adjustment is given by the second derivative in the Taylor expansion of the qth quantile $\alpha_q(L_N)$ of the portfolio loss variable L_N. Here we expanded the portfolio loss, given in equation (9.2), in powers of ε around the conditional mean $\mathbb{E}[L_N|X]$. Equation (9.2) corresponds in the saddlepoint setting to equation (9.25) above. Hence the first derivative with respect to u corresponds to the second derivative with respect to ε. This means that the GA can be identified, in the saddlepoint setting, with the first derivative $\frac{\partial \alpha_q(L_u)}{\partial u}$ of $\alpha_q(L_u)$. Therefore, by approximating the integrals in the above equation around the saddle-point $\hat{t}$, the saddle-point equivalent to the GA formula (9.5), as derived in [102], can be written as

$$\left.\frac{\partial \alpha_q(L_u)}{\partial u}\right|_{u=0} = \frac{1}{\hat{t}} \cdot \left.\frac{\partial}{\partial u}\left(\ln M_{L_u}(s)\right)\right|_{u=0, s=\hat{t}} = \frac{1}{\hat{t} \cdot M_{\mu(X)}(\hat{t})} \cdot \left.\frac{\partial M_{L_u}(s)}{\partial u}\right|_{u=0, s=\hat{t}} \quad (9.26)$$

where the saddle-point $\hat{t}$ is the solution of

$$\left.\frac{\partial M_{\mu(X)}(s)}{\partial s}\right|_{s=\hat{t}} = \alpha_q(\mu(X)) \cdot M_{\mu(X)}(\hat{t}). \quad (9.27)$$

The parameter u can be understood as the amount of concentration risk present in the portfolio. Since $u = 1/N$, the loss variable L_0 can be understood as the portfolio loss of an infinitely granular portfolio and thus (by Theorem 4.1.4) we can identify the distribution of L_0 with that of $\mathbb{E}[L_N|X]$. In [102], the authors show that equation (9.26) can be evaluated further by computing the MGF of L_u which is given by

$$M_{L_u}(s) = \mathbb{E}\left[e^{sL_u}\right] = \int_{-\infty}^{+\infty} e^{s\mu(x)} \cdot e^{s\sqrt{u}\sigma(x)Z} \cdot h(x)dx$$

$$= \int_{-\infty}^{+\infty} e^{s\mu(x)} \left(\int_{-\infty}^{+\infty} \frac{1}{\sqrt{2\pi}} e^{s\sqrt{u}\sigma(x)z} \cdot e^{-z^2/2} dz\right) \cdot h(x)dx$$

$$= \int_{-\infty}^{+\infty} e^{s\mu(x)+u\sigma^2(x)s^2/2} \cdot h(x)dx,$$

h denoting the density function of X. Thus,

$$\left.\frac{\partial M_{L_u}(s)}{\partial u}\right|_{u=0,s=\hat{t}} = \left.\left(\frac{1}{2}s^2 \int_{-\infty}^{+\infty} \sigma^2(x) \cdot e^{s\mu(x)+u\sigma^2(x)s^2/2} \cdot h(x)dx\right)\right|_{u=0,s=\hat{t}}$$

$$= \frac{1}{2}\hat{t}^2 \int_{-\infty}^{+\infty} \sigma^2(x) \cdot e^{\hat{t}\mu(x)} \cdot h(x)dx.$$

Combining this result with equation (9.26), we obtain the following equivalent to the granularity adjustment

$$\left.\frac{\partial \alpha_q(L_u)}{\partial u}\right|_{u=0} = \frac{\hat{t}}{2M_{\mu(X)}(\hat{t})} \int_{-\infty}^{+\infty} \sigma^2(x) \cdot e^{\hat{t}\mu(x)} \cdot h(x)dx. \tag{9.28}$$

[102] discuss the similarities between the saddle-point approximation method and the granularity adjustment and show that they do agree in case of the one-factor CreditRisk$^+$ model. We will present their results in the comparison study of the next Section.

9.4 Discussion and Comparison Study of the Granularity Adjustment Methods

In this section we will first discuss some questions that arise in the granularity adjustment approach. Of course, a first question is whether name concentration is at all relevant. We already presented some results for realistic portfolios in Subsection 9.1.5. However, at this point we will also give an example that demonstrates how drastically name concentration can increase the portfolio VaR. The GA represents, as a model-based approach, a more complicated method for measuring name concentration than, for example, a simple concentration index as discussed in Chapter 8. Hence the question arises, what is the information we gain when using such a complex method? This issue will be discussed and, at least to a certain extend, be answered in Subsection 9.4.2. Other questions that arise within the GA methodology concern the robustness of the GA with respect to regulatory parameters and also with respect to the simplifications made in its derivation. We elaborate on this issue in Subsection 9.4.3.

The most important part of this section, however, concerns the comparison of the different methods to measure name concentration risk. In Subsection 9.4.4 we first compare the GA, as presented in Subsection 9.1, with alternative methods which we did not discuss in detail in these lecture notes but which present related approaches. Moreover, we compare the semi-asymptotic approach of [52] to the GA methodology. Finally, in Subsection 9.4.5 we extensively discuss the similarities between the GA and the saddle-point method. In particular, we prove that in case of the one-factor CreditRisk$^+$ model both approaches agree.

9.4.1 Empirical Relevance of the Granularity Adjustment

We computed the HHI, GA and VaR for an artificially constructed portfolio P which is the most concentrated portfolio that is admissible under the EU *large exposure rules*.[16] To this purpose we divide a total exposure of 6000 into one loan of size 45, 45 loans of size 47 and 32 loans of size 120. We assume a constant PD = 1% and constant ELGD = 45%. For this portfolio we computed an HHI of 0.0156 and a granularity adjustment of $GA = 1.67\%$ of total exposure. The VaR at level 99.9% for portfolio P is approximately 7.72% without GA and 9.39% with GA. Hence the relative add-on for the GA is about 21.61%. This, however, is really the worst case of name concentration that can occur under the large exposure rules. Hence, we can view this add-on for the GA as a sort of upper bound or worst case scenario.[17]

9.4.2 Why a Granularity Adjustment Instead of the HHI?

When discussing granularity adjustment methods a first question that arises is, why should we use any of these GA approaches instead of a simple ad-hoc measure like the HHI? The GA is a much more sophisticated way to measure name concentration than the HHI since it can incorporate obligor specific characteristics like the PD or LGD and, therefore, the GA can reflect the true risk of concentration in a more detailed way. Moreover, the main difficulties when measuring name concentrations with the GA also arise when using a simple ad-hoc measure like the HHI. Here we refer to the problem that name concentration has to be measured on obligor level and not on instrument level, meaning that the exposure to each obligor has to be aggregated before computing the GA or similarly the HHI. Except for this aggregation of data, the GA approach as introduced in [75] uses only information that is already there since it is parameterized to the IRB model. [75] also present an "upper bound" methodology that permits to compute the GA for a part of the portfolio only while cutting back slightly on accuracy. They show that the "upper bound" performs quite well and leads to very good approximations even when including only a small part of the total exposures. Such an "upper bound" can dramatically reduce data requirements in practical applications.

Figure 9.5 shows the dependence of the GA on the HHI for the same data set as described in Subsection 9.1.4. For large portfolios with low HHI, the relationship is almost linear which might suggest an application of the HHI for the measurement of name concentration risk, in particular, because of its simplicity. However, as the figure shows, the GA is much more scattered around the line of linear regression for small and more concentrated portfolios

---

[16] See Directive 93/6/EEC of 15 March 1993 on the capital adequacy of investment firms and credit institutions.
[17] Results for realistic portfolios have already been discussed in Subsection 9.1.5.

with high HHI. These are, however, those portfolios for which an appropriate measurement of concentration risk is more relevant. A reason for this effect is that for smaller portfolios obligor specific default probabilities have a much stronger impact on idiosyncratic risk than for portfolios with low HHI.

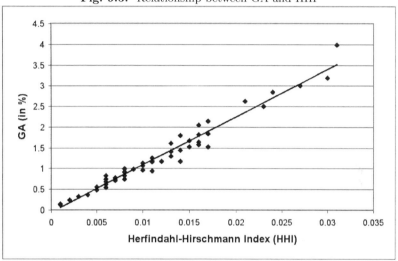

Fig. 9.5. Relationship between GA and HHI

Let us now explain in more detail in which cases the HHI fails to reflect the true risk of name concentration. One can easily see that for constant PDs and LGDs the GA as introduced in Section 9.1 reduces to a multiple of the HHI

$$\overline{GA} = Const(\overline{PD}, \overline{LGD}, q) \cdot \text{HHI},$$

where q denotes the quantile level at which credit VaR is measured.[18] Now, consider a portfolio of one hundred obligors. Obligor 1 has an exposure of $A_1 = 1000$ and $PD_1 = 1\%$, all other obligors have exposure $A = 100$ and $PD = 0.01\%$. The GA with heterogeneous PDs is 7.15% of total exposure whereas the $\overline{GA}$ with weighted average $\overline{PD} = 0.1\%$ is only 1.69% of total exposure. Hence $\overline{GA}$ dramatically underestimates the effect of name concentration in the portfolio due to the assumption of a constant PD.

---

[18] We use the notation ¯ for the simplification of the GA when PDs and LGDs are assumed to be constant.

9.4.3 Accuracy of the Granularity Adjustment and Robustness to Regulatory Parameters

The GA is derived mathematically as a first-order asymptotic approximation for the effect of diversification in large portfolios in the CreditRisk$^+$ model. This raises two possible sources of inaccuracy in the approach. First, as an asymptotic approximation, the GA formula might not work well on small portfolios. Second, the IRB formulae are based on a rather different model of credit risk, so we have a form of "basis risk" (or "model mismatch").

The first issue has been studied in numerical examples by [73] and others. In general, the GA errs on the conservative (i.e., it overstates the effect of granularity), but is quite accurate for modest-sized portfolios of as few as 200 obligors (for a low-quality portfolio) or 500 obligors (for an investment-grade portfolio). Thus, the issue is not a material concern.

The second issue is more difficult. Unfortunately, one cannot test the accuracy of the GA against the IRB model, because it is not possible to construct a "non-asymptotic" generalization of the IRB model. Contrary to the way it is often understood by practitioners and academics alike, the IRB formula is *not* based on a default-mode CreditMetrics model, but rather on a linearized (with respect to maturity) mark-to-market model.[19] The linearization is with respect to the effect of maturity. If the IRB formula had not been linearized this way, then one could implement a single-factor version of the underlying model and quite easily test the accuracy of the GA. However, the linearization makes it impossible to "work backwards" to the underlying model this way. It should be noted as well that the "true" term-structure of capital charges in mark-to-market models tends to be strongly concave, so the linearization was not at all a minor adjustment. In order to minimize the potential inaccuracy due to the differences between the mark-to-market basis of the IRB and the default-mode origins of the GA, [75] constructed the GA formula using IRB inputs ($\mathcal{K}$, in particular) that have been maturity-adjusted. Thus, the output of the GA formula is implicitly maturity-adjusted, albeit in a not very rigorous manner.

As it is not possible to assess directly the accuracy of the GA against the IRB model, we must reinterpret questions of accuracy as questions concerning the robustness of the GA to its parameterization. One question is whether the simplified form (9.12) of the formula might be inaccurate against the exact form (9.10). The simplified GA errs on the side of lenience, but [75] find that the error is very small except for low-quality portfolios of very high concentration. Such portfolios would come under supervisory scrutiny even in the absence of a GA.

---

[19] Actual calibration was made to KMV Portfolio Manager, but with extensive comparison against alternative approaches.

Another question concerns the sensitivity of the GA to regulatory parameters, especially the CreditRisk$^+$ parameter ξ and the specification of the variance of LGD. The sensitivity of the GA to the variance parameter ξ of the systematic factor has been explored in Subsection 9.1.5. We saw that the GA is strictly increasing in ξ, and that the degree of sensitivity of the absolute magnitude of the GA is not negligible. The relative magnitude of the GA across bank portfolios, however, is much less sensitive to the parameter ξ such that a proper function of the GA as a supervisory tool does not materially depend on the precision with which ξ is calibrated.

9.4.4 Comparison of Granularity Adjustment with Other Model-Based Approaches

We saw in Subsection 9.4.2 that simple ad-hoc measures like the HHI are, in general, no real alternatives to the GA methodology. Another approach, due to [120], lies somewhere between ad-hoc and model-based. In this method, one augments systematic risk (by increasing the factor loading) in order to compensate for ignoring the idiosyncratic risk. The trouble is that systematic and idiosyncratic risk have very different distribution shapes. This method is known to perform quite poorly in practice, and therefore has not been presented in this chapter.

Much closer to the proposal of [75] in spirit and methodology is the approach of [52]. The latter offer a granularity adjustment based on a one-factor default-mode CreditMetrics model, which has the advantage of relative proximity of the model underpinning the IRB formula. As discussed earlier in this section, however, we believe this advantage to be more in appearance than in substance because of the importance of maturity considerations in the IRB model. As a mark-to-market extension of the [52] GA appears to be intractable, maturity considerations would need to be introduced indirectly (as in the proposal of [75]) through the inputs. Reparameterization of the GA approach in [52] along these lines is feasible in principle, but would lead to a rather more complicated formula with more inputs than the CreditRisk$^+$-based GA of [75].

The Semi-Asymptotic approach of [52] relies on a limiting procedure that is applied only to a part of the portfolio. In the limit as the number N of loans in the portfolio tends to infinity, the squared exposure shares of all but one obligor vanish and only a single exposure share is assumed to remain constant. The granularity adjustment method is based on exactly this assumption of vanishing squared exposure shares. However, for highly concentrated portfolios with very large exposures this assumption is questionable such that the semi-asymptotic approach might be preferable in this situation. The numerical results in [52], however, show that the GA methodology yields very precise approximations to the true portfolio GA as long as the relative weight of the largest loan in the portfolio (the one which is assumed to have constant relative weight in the semi-asymptotic approach) is less than 5%. Moreover, the semi-asymptotic approach can also lead to counter-intuitive results when the PD of the new loan is very small. This is, of course, a major drawback of the method as the largest loans in a portfolio usually are of very high credit quality. [52] mention that this shortcoming of the semi-asymptotic approach can be circumvented when using a different, and in particular, a coherent risk measure as, for example, Expected Shortfall.

An alternative that has not been much studied in the literature is the saddle-point based method of [102] which we presented in Section 9.3. Results in that paper suggest that it would be quite similar to the GA in performance and pose a similar tradeoff between fidelity to the IRB model and analytical tractability. Indeed, it is not at all likely that the saddle-point GA would yield a closed-form solution for any industry credit risk model other than CreditRisk+. In that case, however, both methods agree as we will demonstrate in the next subsection.

9.4.5 Agreement of Granularity Adjustment and Saddle-Point Approximation Method in the CreditRisk+ Model

In CreditRisk+ the systematic risk factor X is $\Gamma(\alpha, \beta)$-distributed, i.e. the mean of X equals $\alpha\beta$ and its variance is $\alpha\beta^2$. Default probabilities in CreditRisk+ are proportional to the systematic risk factor X. The conditional mean $\mu(X)$ and the conditional variance $\sigma^2(X)$ in CreditRisk+ can be expressed in terms of the scaled variable $X/\alpha\beta$ as

$$\mu(X) = \sum_{n=1}^{N} s_n \cdot \text{ELGD}_n \cdot \text{PD}_n \cdot X$$

$$\sigma^2(X) = \left(\sum_{n=1}^{N} s_n^2 \cdot \text{ELGD}_n^2 \cdot \text{PD}_n \right) \frac{X}{\alpha\beta},$$

respectively, where s_n is the exposure share of obligor n and PD_n its (unconditional) default probability.[20] Inserting this in formula (9.28), we obtain for the granularity adjustment

$$\left. \frac{\partial a_q(L_u)}{\partial u} \right|_{u=0} = \frac{\hat{t} \sum_{n=1}^{N} s_n^2 \cdot \text{ELGD}_n^2 \cdot \text{PD}_n}{2\alpha\beta M_{\mu(X)}(\hat{t})} \int_{-\infty}^{+\infty} x \cdot e^{\hat{t}\mu(x)} \cdot h(x) dx. \quad (9.29)$$

From equation (9.27) we know that the saddle-point $\hat{t}$ is the solution of the equation

---

[20] This corresponds to the formulae for the conditional mean and variance in Section 9.1.3 when neglecting squares of default probabilities and when assuming a factor loading of one and a non-random LGD.

9.4 Discussion and Comparison Study

$$\alpha_q(\mu(X)) = \frac{1}{M_{\mu(X)}(\hat{t})} \cdot \frac{\partial M_{\mu(X)}(s)}{\partial s}\bigg|_{s=\hat{t}}$$

$$= \frac{1}{M_{\mu(X)}(\hat{t})} \int_{-\infty}^{+\infty} \mu(x) \cdot e^{\hat{t}\mu(x)} \cdot h(x) dx$$

$$= \frac{\sum_{n=1}^{N} s_n \cdot \text{ELGD}_n \cdot \text{PD}_n}{M_{\mu(X)}(\hat{t})} \int_{-\infty}^{+\infty} x \cdot e^{\hat{t}\mu(x)} \cdot h(x) dx.$$

Inserting in equation (9.29), we obtain

$$\frac{\partial \alpha_q(L_u)}{\partial u}\bigg|_{u=0} = \frac{\hat{t} \cdot \alpha_q(\mu(X))}{2\alpha\beta \left(\sum_{n=1}^{N} s_n \, \text{ELGD}_n \, \text{PD}_n\right)} \sum_{n=1}^{N} s_n^2 \cdot \text{ELGD}_n^2 \cdot \text{PD}_n . \tag{9.30}$$

Continuing the above calculation yields

$$\alpha_q(\mu(X)) = \frac{\sum_{n=1}^{N} s_n \, \text{ELGD}_n \, \text{PD}_n}{M_{\mu(X)}(\hat{t})} \int_{-\infty}^{+\infty} x \cdot e^{\hat{t}\mu(x)} \cdot h(x) dx$$

$$= \frac{\sum_{n=1}^{N} s_n \, \text{ELGD}_n \, \text{PD}_n}{M_{\mu(X)}(\hat{t})} \cdot \frac{\partial M_X(s)}{\partial s}\bigg|_{s=\hat{t}\cdot \sum_{n=1}^{N} s_n \, \text{ELGD}_n \, \text{PD}_n}. \tag{9.31}$$

Note that

$$M_{\mu(X)}(\hat{t}) = \mathbb{E}\left[e^{\hat{t}\mu(X)}\right] = \mathbb{E}\left[e^{\hat{t}\cdot(\sum_{n=1}^{N} s_n \cdot \text{ELGD}_n \cdot \text{PD}_n)X}\right]$$

$$= M_X\left(\hat{t} \cdot \sum_{n=1}^{N} s_n \cdot \text{ELGD}_n \cdot \text{PD}_n\right).$$

Since the systematic risk factor X is $\Gamma(\alpha, \beta)$-distributed in the CreditRisk$^+$ model, we have

$$\frac{\partial M_X(s)}{\partial s} = \frac{-\alpha}{s - 1/\beta} \cdot M_X(s).$$

Due to equation (9.31), this implies that the saddle-point $\hat{t}$ is the solution of

$$\alpha_q(\mu(X)) = \frac{-\alpha \sum_{n=1}^{N} s_n \, \text{ELGD}_n \, \text{PD}_n}{\hat{t} \cdot \sum_{n=1}^{N} s_n \, \text{ELGD}_n \, \text{PD}_n - 1/\beta} \cdot \frac{M_X\left(\hat{t} \cdot \sum_{n=1}^{N} s_n \, \text{ELGD}_n \, \text{PD}_n\right)}{M_{\mu(X)}(\hat{t})}$$

$$= \frac{-\alpha \sum_{n=1}^{N} s_n \, \text{ELGD}_n \, \text{PD}_n}{\hat{t} \cdot \sum_{n=1}^{N} s_n \, \text{ELGD}_n \, \text{PD}_n - 1/\beta}$$

which is equivalent to

$$\hat{t} = \frac{1}{\beta \cdot \sum_{n=1}^{N} s_n \, \text{ELGD}_n \, \text{PD}_n} - \frac{\alpha}{\alpha_q(\mu(X))}. \tag{9.32}$$

9 Name Concentration

Inserting in equation (9.30) and using $\alpha_q(\mu(X)) = \mu(\alpha_q(X))$ (compare Theorem 4.1.6), we obtain

$$\frac{\partial \alpha_q(L_u)}{\partial u}\bigg|_{u=0} = \frac{1}{2\alpha\beta} \cdot \left(\frac{1}{\beta \cdot \sum_{n=1}^{N} s_n \, \text{ELGD}_n \, \text{PD}_n} - \frac{\alpha}{\alpha_q(\mu(X))}\right) \cdot \alpha_q(\mu(X))$$

$$\cdot \left(\sum_{n=1}^{N} s_n \, \text{ELGD}_n \, \text{PD}_n\right)^{-1} \cdot \left(\sum_{n=1}^{N} s_n^2 \, \text{ELGD}_n^2 \, \text{PD}_n\right)$$

$$= \frac{1}{2\alpha\beta^2} \cdot \left(\frac{\mu(\alpha_q(X))}{\sum_{n=1}^{N} s_n \, \text{ELGD}_n \, \text{PD}_n} - \alpha\beta\right)$$

$$\cdot \left(\sum_{n=1}^{N} s_n \, \text{ELGD}_n \, \text{PD}_n\right)^{-1} \cdot \left(\sum_{n=1}^{N} s_n^2 \, \text{ELGD}_n^2 \, \text{PD}_n\right)$$

$$= \frac{\sum_{n=1}^{N} s_n^2 \, \text{ELGD}_n^2 \, \text{PD}_n}{2\alpha\beta^2 \cdot \left(\sum_{n=1}^{N} s_n \, \text{ELGD}_n \, \text{PD}_n\right)} \cdot (\alpha_q(X) - \alpha\beta).$$

This corresponds to the GA given by equation (9.10) when ignoring squares of PDs and assuming the loss given defaults to be constant equal to the average loss given defaults ELGD_n for all $n = 1, \ldots, N$. To see this correspondence more clearly, we adopt the $\mathcal{K}_n$ and $\mathcal{R}_n$ notation of Subsection 9.1.3. Moreover, in analogy to Subsection 9.1.3, we use $\alpha\beta = 1$ and $1/\beta = \xi$. Then we obtain for the granularity adjustment, as computed by the saddle-point approximation method,

$$\frac{\partial \alpha_q(L_u)}{\partial u}\bigg|_{u=0} = \frac{1}{2\mathcal{K}^*} \sum_{n=1}^{N} s_n^2 \left[(\alpha_q(X) - 1) \cdot \xi \cdot \left(\text{ELGD}_n^2 \, \text{PD}_n \, (\alpha_q(X) - 1)\right)\right].$$

Note that the remaining terms in equation (9.10) vanish in this simplified CreditRisk$^+$ setting as the factor loading to the single systematic risk factor is $w_n = 1$ for all n in this setting and $\text{VLGD}_n = 0$. This shows that both methods agree in the case of the single factor CreditRisk$^+$ model.

10
Sector Concentration

Although the importance of an appropriate methodology to quantify sector concentration risk is broadly recognized, the question how to measure sector concentration risk has not been answered yet. The occurrence of sector concentration directly results in violating the assumptions of the analytical Merton-type model, which constitutes the core under Pillar I of the Basel II framework for quantifying credit risk. In order to ensure analytical tractability of the model, the computation of regulatory capital is based on the so-called Asymptotic Single Risk Factor (ASRF) framework developed by the BCBS and, in particular, by [72]. However, the ASRF model assumes that all obligors depend on the same single systematic risk factor. Thus it measures only systematic risk and fails in detecting exposure concentrations as well as in recognizing diversification effects. The assumption of a single risk factor implies a uniform correlation among obligors and thereby can lead to an over- or underestimation of risk for portfolios with unequally distributed sector structure. Hence, to account for sector concentration risk a more general framework has to be considered.

Prior to any model for sector concentration, however, it is necessary to find a meaningful definition of sectors such that the asset correlation among obligors in the same sector is very high and among different sectors is very low. The number of sectors taken into account in a specific model thereby is strongly determined by the available data and a stable estimation of correlations. The problem of finding a meaningful definition of sectors is a problem of its own and will not be addressed in these lecture notes. However, one should always keep in mind that the estimation of correlation parameters can lead to substantial problems and, in practice, it is often one of the main problems risk managers have to face.

As in the case of name concentration, there are a lot of ad-hoc measures which can also be applied to measure sector concentration when taking into account the already mentioned shortcomings. For a more accurate measurement of sector concentration one can apply certain multi-factor models which have become increasingly popular in credit risk modeling. They base on the principle that defaults of individual obligors in the portfolio are independent conditional on a set of common systematic risk factors describing the state of the economy. Representatives of these models have been introduced in Chapter 3 in the example of the multi-factor Merton model and in Chapter 6 in the example of the multi-factor CreditRisk$^+$ model, where each geographical or industrial sector can be represented by a certain risk factor. The extend to which a bank is concentrated in sectors then depends on the distribution of exposures among sectors and the correlation between the individual sectors. Thus, the determination of correlations between the single risk factors is essential in such a multi-factor model and has a strong impact on the total portfolio risk.

In case of the multi-factor Merton model the asset-returns of the individual borrowers are determined by K normally distributed systematic risk factors and an obligor-specific idiosyncratic shock which is also assumed to be normally distributed. Due to the conditional independence framework it is possible to derive a formula for the portfolio loss variable. The computation of quantiles, as for example Value-at-Risk (VaR), then requires a simulation of the portfolio loss distribution. This is usually achieved by Monte Carlo methods. Even though these simulation-based models are widely used, all these methods have in common that they are quite time-consuming and, more importantly, portfolio dependent. This means that, whenever a new loan is added to the portfolio, the loss distribution has to be simulated anew. Hence it is not possible within these approaches to simply compute the contribution of the new loan to portfolio VaR and then combine with the former portfolio VaR. However, in credit portfolio management as well as for supervisory applications fast analytical methods are often preferable. The granularity adjustment method presented in Chapter 9 represents an example of such a technique for the case of the one-factor CreditRisk$^+$ model. We devote this chapter to different analytical methods for measuring sector concentration.

In Section 10.1 we present an analytical approximation to Value-at-Risk which leads to a multi-factor adjustment similar in spirit to the granularity adjustment of Section 9.1. Afterwards, in Section 10.2 we discuss a model which is, in a sense, semi-analytic as it is characterized by a numerically estimated scaling factor which is then applied to the economic capital required in the ASRF model. Both approaches are based on the multi-factor Merton model, presented in Section 3.2.

10.1 Analytical Approximation Models

In this section we present a model that describes an extension of the idea of the granularity adjustment to the multi-factor situation. [109] derive an analytical approximation for portfolio Value-at-Risk and Expected Shortfall within the multi-factor Merton framework. They use the one-factor limiting loss distribution of the ASRF model as an approximation to the multi-factor loss distribution. Similarly to the granularity adjustment approach, the methodology is based on a 2nd order Taylor expansion of portfolio VaR around the limiting portfolio loss variable. The resulting formula consists of two analytical expressions. First of all, one has to consider the difference between the multi-factor and the one-factor limiting loss distribution. Thus, the first component is a *multi-factor adjustment* that takes care of the multiple factors that determine the risk of the portfolio. Taking this difference into account is of great relevance because single-factor models cannot reflect segmentation effects. For example, if different industries and geographic regions underlie different cycles, they should be reflected by distinct systematic factors.

Moreover, the loss distribution in the ASRF model is derived under the assumption of an infinitely fine-grained portfolio of loans. Hence the second component in the analytical approximation to VaR represents a correction for neglecting granularity in the portfolio. Thus it can be understood as a *granularity adjustment*. This aspect is of importance since real-world portfolios are rarely infinitely fine-grained. We discussed this issue already in Chapter 9. One can easily see that in the single-factor situation under which we derived the GA in Section 9.1, the analytical approximation to VaR as presented in [109] coincides with the GA.

10.1.1 Analytical Approximation for Value-at-Risk

We will first review the method of [109] to derive an analytic approximation for the qth quantile $\alpha_q(L_N)$ of the portfolio loss variable L_N within the multi-factor Merton model presented in Section 3.2. For simplicity of notation we will, in this section, skip the index N in the portfolio loss $L = L_N$, as the number N of borrowers in the portfolio will be fix throughout the whole section. Moreover, we adopt the notation of Section 3.2.

Since we cannot calculate $\alpha_q(L)$ analytically in this multi-factor framework, [109] adopt the approach used in the derivation of the granularity adjustment. First, they construct a random variable $\bar{L}$ for which we can compute the qth quantile, $\alpha_q(\bar{L})$, and which is close to $\alpha_q(L)$. Afterwards, [109] apply a 2nd order Taylor expansion to approximate the quantile $\alpha_q(L)$ of the portfolio loss by that of $\bar{L}$. This approximation can then be split in a multi-factor adjustment term and a granularity adjustment term.

Let us start with the derivation of the variable $\bar{L}$. Since one can derive the loss distribution of equation (3.10) analytically in the one-factor Merton framework, it is obvious to define $\bar{L}$ as the limiting loss distribution for the same portfolio but in the one-factor setting, i.e.

$$\bar{L} = \mu(\bar{X}) = \sum_{n=1}^{N} s_n \cdot \mathrm{ELGD}_n \cdot \mathrm{PD}_n(\bar{X}), \tag{10.1}$$

where $\bar{X}$ is the single systematic risk factor having a standard normal distribution. Moreover, in the single-factor Merton model, the probability of default of borrower n conditional on $\bar{X} = x$ can be computed as[1]

$$\mathrm{PD}_n(x) = \Phi\left(\frac{\Phi^{-1}(\mathrm{PD}_n) - w_n x}{\sqrt{1-w_n^2}}\right),$$

where w_n represents the effective factor loading of borrower n. Since $\mu(\bar{X})$ is a deterministic monotonically decreasing function of $\bar{X}$, we can calculate $\alpha_q(\bar{L})$ as follows[2]

$$\alpha_q(\bar{L}) = \mu(\alpha_{1-q}(\bar{X})) = \mu(\Phi^{-1}(1-q)). \tag{10.2}$$

This equation determines the loss if the risk factor is in the $(1-q)^{\mathrm{th}}$ percentile. For instance, if q is close to one, this equation reflects the loss given that the state of the economy described by the single systematic risk factor is bad.

To relate the random variable $\bar{L}$ to the portfolio loss L, we need a relationship between the effective systematic factor $\bar{X}$ and the independent risk factors $X_1, \ldots, X_K$. Assume

$$\bar{X} = \sum_{k=1}^{K} b_k X_k,$$

where $\sum_{k=1}^{K} b_k^2 = 1$ to preserve unit variance of $\bar{X}$. In order to determine $\bar{L}$ completely, one has to specify the factor loadings w_n and the coefficients b_k. In analogy to the granularity adjustment approach, an intuitive requirement is to assume that $\bar{L}$ equals the expected loss conditional on $\bar{X}$, that is $\bar{L} = \mathbb{E}[L|\bar{X}]$. To calculate $\mathbb{E}[L|\bar{X}]$ we represent the composite risk factor Y_n for borrower n as

$$Y_n = \varrho_n \cdot \bar{X} + \sqrt{1-\varrho_n^2} \cdot \eta_n, \tag{10.3}$$

where η_n is a standard normal variable independent of $\bar{X}$ and

$$\varrho_n = \mathrm{Corr}(Y_n, \bar{X}) = \sum_{k=1}^{K} \alpha_{n,k} b_k$$

represents the correlation between Y_n and $\bar{X}$.[3] Equation (10.3) points out that borrower n's risk factors can be decomposed into a single (or global) risk

---

[1] Compare equation (3.8).
[2] Compare Proposition 2.3.3.
[3] Compare equation (3.7).

factor $\bar{X}$ and a specific component η_n of borrower n. Due to this definition one can rewrite equation (3.6) as[4]

$$r_n = \beta_n \varrho_n \cdot \bar{X} + \sqrt{1 - (\beta_n \varrho_n)^2} \cdot \varepsilon_n, \qquad (10.4)$$

where ε_n is a standard normal variable independent of $\bar{X}$. Since

$$\mathbb{E}[r_n | \bar{X}] = \beta_n \varrho_n \bar{X} \quad \text{and} \quad \mathbb{V}[r_n | \bar{X}] = 1 - \beta_n^2 \varrho_n^2,$$

the conditional expectation of L is

$$\mathbb{E}[L | \bar{X}] = \sum_{n=1}^{N} s_n \cdot \mathrm{ELGD}_n \cdot \Phi\left(\frac{\Phi^{-1}(\mathrm{PD}_n) - \beta_n \varrho_n \cdot \bar{X}}{\sqrt{1 - \beta_n^2 \varrho_n^2}} \right). \qquad (10.5)$$

By comparing the equations (10.1) and (10.5) for $\bar{L}$ and $\mathbb{E}[L|\bar{X}]$, we obtain that $\bar{L} = \mathbb{E}[L|\bar{X}]$ for any portfolio decomposition if and only if the following restriction for the effective factor loadings holds

$$\omega_n = \beta_n \varrho_n = \beta_n \sum_{k=1}^{K} \alpha_{n,k} b_k. \qquad (10.6)$$

In the following we will always assume the relation (10.6) to hold.

Remark 10.1.1 As ω_n depends on b_k and $\alpha_{n,k}$, either have to be specified. Since $\alpha_q(\bar{L})$ depends on b_k, the latter should be chosen such that the difference between $\alpha_q(\bar{L})$ and $\alpha_q(L)$ is minimized in order to receive a good approximation. An intuitive way is to maximize the correlation between $\bar{X}$ and Y_n for all n. A correlation close to one emphasizes that the single-factor and the multi-factor model are influenced by similar information leading to a small difference between $\alpha_q(\bar{L})$ and $\alpha_q(L)$. This leads to the following maximization problem to obtain b_k

$$\max_{b_1, \ldots, b_K} \left(\sum_{n=1}^{N} d_n \sum_{k=1}^{K} \alpha_{n,k} b_k \right) \quad \text{such that} \quad \sum_{k=1}^{K} b_k^2 = 1.$$

[109] show that the solution to this problem is given by

$$b_k = \sum_{n=1}^{N} \frac{d_n}{\lambda} \cdot \alpha_{nk},$$

where λ is the Lagrange multiplier such that $\sum_{k=1}^{K} b_k^2 = 1$. This maximization problem, however, produces another undefined variable, called d_n. After trying different specifications, [109] find out that

---

[4] If we plug equation (10.3) into (3.6), we receive $\mathbb{E}[r_n] = 0$ and $\mathbb{V}[r_n] = 1$. Moreover, the log-returns r_n remain normally distributed, which is consistent with the new formula for r_n.

$$d_n = s_n \cdot \text{ELGD}_n \cdot \Phi\left(\frac{\Phi^{-1}(\text{PD}_n) - \beta_n \Phi^{-1}(q)}{\sqrt{1-\beta_n^2}}\right)$$

is one of the best-performing choices.

Having constructed the variable $\bar{L}$, [109] perturb the real portfolio loss variable L by a perturbation $U := L - \bar{L}$. The perturbed variable L_ε is defined as

$$L_\varepsilon = \bar{L} + \varepsilon U = \bar{L} + \varepsilon(L - \bar{L}).$$

Note that for $\varepsilon = 1$ we have $L_\varepsilon = L$. Similarly to the derivation of the granularity adjustment, [109] expand the qth quantile $\alpha_q(L_\varepsilon)$ of the perturbed portfolio loss L_ε in powers of ε around $\bar{L}$ and evaluate at $\varepsilon = 1$. This second-order Taylor expansion yields

$$\alpha_q(L) = \alpha_q(\bar{L}) + \left.\frac{d\alpha_q(L_\varepsilon)}{d\varepsilon}\right|_{\varepsilon=0} + \frac{1}{2}\left.\frac{d^2\alpha_q(L_\varepsilon)}{d\varepsilon^2}\right|_{\varepsilon=0} + \mathcal{O}(\varepsilon^3). \tag{10.7}$$

From equation (10.2) we know that the zeroth-order approximation $\alpha_q(\bar{L})$ in the expansion (10.7) is simply given by $\alpha_q(\bar{L}) = \mu(\Phi^{-1}(1-q))$. The derivatives in equation (10.7) can be calculated by the results in [76].

We first turn to the approximation of the first-order term. Note that $L_\varepsilon = \bar{L}$ for $\varepsilon = 0$. As proved by [76] and restated in Lemma 9.1.1, the first derivative can be written as

$$\left.\frac{d\alpha_q(L_\varepsilon)}{d\varepsilon}\right|_{\varepsilon=0} = \mathbb{E}[U|\bar{L} = \alpha_q(\bar{L})] = \mathbb{E}[U|\bar{X} = \Phi^{-1}(1-q)] \tag{10.8}$$

by equations (10.1) and (10.2). Since the expected value of the perturbation conditional on the risk factors is zero, i.e. $\mathbb{E}[U|\bar{X}] = 0$, the first-order term in the Taylor series vanishes for any level q.[5] Hence the remaining task is to derive an explicit algebraic expression for the second-order term. As proved in Lemma 3 of [76], the second derivative can be expressed as

$$\left.\frac{d^2\alpha_q(L_\varepsilon)}{d\varepsilon^2}\right|_{\varepsilon=0} = -\frac{1}{f_{\bar{L}}(\alpha_q(\bar{L}))} \cdot \left.\frac{d}{dl}\left(f_{\bar{L}}(l) \cdot \mathbb{V}[U|\bar{L}=l]\right)\right|_{l=\alpha_q(\bar{L})}, \tag{10.9}$$

where $f_{\bar{L}}(l)$ is the probability density function of $\bar{L}$. Recall that $\bar{L} = \mu(\bar{X})$ is a deterministic and monotonically decreasing function of $\bar{X}$. This implies, in particular, that the conditional variance of L is the same as the conditional variance of U given $\bar{X} = x$, i.e.

$$\sigma^2(x) := \mathbb{V}[L|\bar{X} = x] = \mathbb{V}[U|\bar{X} = x].$$

By substitution of $l = \mu(x)$ in equation (10.9), we can rewrite the above equation as

---

[5] Compare Section 9.1.

$$\left.\frac{d^2\alpha_q(L_\varepsilon)}{d\varepsilon^2}\right|_{\varepsilon=0} = -\frac{1}{\phi(\alpha_{1-q}(\bar{X}))} \cdot \frac{d}{dx}\left(\phi(x) \cdot \frac{\sigma^2(x)}{\mu'(x)}\right)\Bigg|_{x=\alpha_{1-q}(\bar{X})}, \quad (10.10)$$

where $\phi(y)$ is the density of the effective risk factor $\bar{X}$ which is standard normal. Taking the derivative in equation (10.10) and inserting in equation (10.7), we can write the correction term as

$$\Delta\alpha_q = \alpha_q(L) - \alpha_q(\bar{L})$$

$$= -\frac{1}{2\mu'(\alpha_{1-q}(\bar{X}))} \cdot \left[\frac{d}{dx}\sigma^2(x) - \sigma^2(x) \cdot \left(\frac{\mu''(x)}{\mu'(x)} + x\right)\right]_{x=\alpha_{1-q}(\bar{X})} \quad (10.11)$$

where we used $\phi'(x)/\phi(x) = -x$. Moreover, we used that the first-order term in the Taylor expansion (10.7) is zero. This term represents the adjustment which involves two important components. First, it considers the adjustment between the multi-factor and the single-factor limiting loss distribution. Second, it includes the correction between a finite and an infinite number of loans.

To derive an explicit adjustment $\Delta\alpha_q$ we have to determine the conditional mean and variance of L given $\bar{X} = x$ and their derivatives. The first and second derivative of $\mu(x)$ can be easily obtained from equation (10.1). Therefore, it only remains to determine $\sigma^2(x)$ and $\frac{d}{dx}\sigma^2(x)$. Since the asset returns are independent conditional on $\{X_1,\ldots,X_K\}$, we can decompose the variance $\sigma^2(x)$ into the sum of the systematic and the idiosyncratic risk[6]

$$\sigma^2(x) = \underbrace{\mathbb{V}\left[\mathbb{E}[L|\{X_1,\ldots,X_K\}]|\bar{X}=x\right]}_{\sigma^2_\infty(x)} + \underbrace{\mathbb{E}\left[\mathbb{V}[L|\{X_1,\ldots,X_K\}]|\bar{X}=x\right]}_{\sigma^2_{GA}(x)}. \quad (10.12)$$

The first term on the right-hand side is the variance of the limiting loss distribution (3.10) conditional on the effective risk factor $\bar{X} = x$. It represents an adjustment for the difference between the multi-factor and the single-factor limiting loss distribution. This statement is accentuated by the fact that the first variance term disappears if the single systematic factor $\bar{X}$ is equal to the independent factors $\{X_1,\ldots,X_K\}$, which is the case in the GA of Chapter 9. The following result is due to [109].

Theorem 10.1.2 *The conditional variance term σ^2_∞ is given by*

$$\sigma^2_\infty(x) = \sum_{n=1}^{N}\sum_{m=1}^{N} s_n s_m \cdot \mathrm{ELGD}_n \cdot \mathrm{ELGD}_m \cdot$$
$$\cdot \left[\Phi_2\left(\Phi^{-1}(\mathrm{PD}_n(x)), \Phi^{-1}(\mathrm{PD}_m(x)), \varrho^{\bar{X}}_{nm}\right) - \mathrm{PD}_n(x)\,\mathrm{PD}_m(x)\right], \quad (10.13)$$

---

[6] See Berger and Casella (2002), p. 167.

where $\Phi_2(.,.,.)$ is a bivariate normal distribution and $\varrho_{nm}^{\bar{X}}$ is the asset correlation between asset n and asset m conditional on $\bar{X}$. Moreover, the derivative of σ_∞^2 is given by

$$\frac{d}{dx}\sigma_\infty^2(x) = 2 \sum_{n=1}^N \sum_{m=1}^N s_n s_m \cdot \mathrm{ELGD}_n \cdot \mathrm{ELGD}_m \cdot \mathrm{PD}_n'(x) \cdot \\ \cdot \left[\Phi\left(\frac{\Phi^{-1}(\mathrm{PD}_m(x)) - \varrho_{nm}^{\bar{X}} \cdot \Phi^{-1}(\mathrm{PD}_n(x))}{\sqrt{1 - (\varrho_{nm}^{\bar{X}})^2}} \right) - \mathrm{PD}_m(x) \right]. \tag{10.14}$$

where $\mathrm{PD}_n'(x)$ denotes the derivative of $\mathrm{PD}(x)$ with respect to x.

Remark 10.1.3 The conditional correlation can be derived as follows. By exploiting the definitions of Y_n and $\bar{X}$ we can rewrite equation (3.6) as

$$r_n = \omega_n \cdot \bar{X} + \sum_{k=1}^K (\beta_n \alpha_{nk} - \omega_n b_k) \cdot X_k + \sqrt{1 - \beta_n^2} \cdot \varepsilon_n.$$

Now, we calculate the asset correlation conditional on $\bar{X}$ by applying the relation $\omega_n = \beta_n \varrho_n$ and the restriction $\sum_{k=1}^K b_k^2 = 1$. Then we obtain

$$\varrho_{nm}^{\bar{X}} = \frac{\beta_n \beta_m \sum_{k=1}^K \alpha_{n,k} \alpha_{m,k} - \omega_n \omega_m}{\sqrt{(1 - \omega_n^2)(1 - \omega_m^2)}}.$$

Proof of Theorem 10.1.2. Denote the limiting loss distribution of the multifactor Merton model by $L^\infty := \mathbb{E}[L|\{X_1, \ldots, X_K\}]$. From equation (3.9) we obtain

$$L^\infty = \sum_{n=1}^N s_n \cdot \mathrm{ELGD}_n \cdot \mathbf{1}_{\{r_n \leq \Phi^{-1}(\mathrm{PD}_n)\}},$$

where the asset log-returns r_n are normally distributed. Define the obligor specific loss variables by

$$L_n^\infty = \mathrm{ELGD}_n \cdot \mathbb{E}\left[\mathbf{1}_{\{r_n \leq \Phi^{-1}(\mathrm{PD}_n)\}} | \{X_1, \ldots, X_K\}\right].$$

Then

$$\mathrm{Cov}[s_n L_n^\infty, s_m L_m^\infty | \bar{X}]$$
$$= \mathbb{E}[s_n s_m \cdot L_n^\infty \cdot L_m^\infty | \bar{X}] - \mathbb{E}[s_n L_n^\infty | \bar{X}] \cdot \mathbb{E}[s_m L_m^\infty | \bar{X}]$$
$$= s_n s_m \cdot \mathrm{ELGD}_n \cdot \mathrm{ELGD}_m \cdot \Big(\mathbb{E}\left[\mathbf{1}_{\{r_n \leq \Phi^{-1}(\mathrm{PD}_n)\}} \cdot \mathbf{1}_{\{r_m \leq \Phi^{-1}(\mathrm{PD}_m)\}} | \bar{X}\right]$$
$$- \mathbb{E}\left[\mathbf{1}_{\{r_n \leq \Phi^{-1}(\mathrm{PD}_n)\}} | \bar{X}\right] \cdot \mathbb{E}\left[\mathbf{1}_{\{r_m \leq \Phi^{-1}(\mathrm{PD}_m)\}} | \bar{X}\right] \Big).$$

Recall that, conditional on $\bar{X}$, the asset log-returns r_n and r_m of obligors n and m are normally distributed with correlation $\varrho_{nm}^{\bar{X}}$ such that

$$\mathbb{E}\left[\mathbf{1}_{\{r_n \leq \Phi^{-1}(\mathrm{PD}_n)\}} \cdot \mathbf{1}_{\{r_m \leq \Phi^{-1}(\mathrm{PD}_m)\}} | \bar{X}\right]$$
$$= \Phi_2\left(\Phi^{-1}(\mathrm{PD}_n(x)), \Phi^{-1}(\mathrm{PD}_m(x)), \varrho_{nm}^{\bar{X}}\right),$$

which implies the first result.

Hence it remains to prove the equation for the derivative of σ_∞^2. Recall that a bivariate normal distribution with mean zero, standard deviation equal to 1 and a correlation matrix ϱ_{12} can be written as

$$\Phi_2(x_1, x_2, \varrho_{12}) = \int_{-\infty}^{x_1} \int_{-\infty}^{x_2} \frac{1}{2\pi\sqrt{1-\varrho_{12}^2}} \cdot \exp\left(-\frac{x^2 + y^2 - 2\varrho_{12}xy}{2(1-\varrho_{12}^2)}\right) dx\,dy.$$

Thus we have

$$\frac{\partial}{\partial x_1} \Phi_2(x_1, x_2, \varrho_{12})$$
$$= \int_{-\infty}^{x_2} \frac{1}{2\pi\sqrt{1-\varrho_{12}^2}} \cdot \exp\left(-\frac{x_1^2 + y^2 - 2\varrho_{12}\cdot x_1 y}{2(1-\varrho_{12}^2)}\right) dy$$
$$= \int_{-\infty}^{x_2} \frac{1}{2\pi\sqrt{1-\varrho_{12}^2}} \cdot \exp\left(-\left[\frac{x_1^2}{2} + \frac{(y - \varrho_{12}\cdot x_1)^2}{2(1-\varrho_{12}^2)}\right]\right) dy$$
$$= \Phi\left(\frac{(x_2 - \varrho_{12} x_1)}{\sqrt{1-\varrho_{12}^2}}\right) \cdot \phi(x_1).$$

Hence we obtain

$$\frac{\partial}{\partial x}\Phi_2(\Phi^{-1}(\mathrm{PD}_n(x)), \Phi^{-1}(\mathrm{PD}_m(x)), \varrho_{nm}^{\bar{X}})$$
$$= \Phi\left(\frac{\Phi^{-1}(\mathrm{PD}_m(x)) - \varrho_{nm}^{\bar{X}} \cdot \Phi^{-1}(\mathrm{PD}_n(x))}{\sqrt{1-(\varrho_{nm}^{\bar{X}})^2}}\right)$$
$$\cdot \phi(\Phi^{-1}(\mathrm{PD}_n(x))) \cdot \frac{d}{dx}[\Phi^{-1}(\mathrm{PD}_n(x))]$$
$$= \Phi\left(\frac{\Phi^{-1}(\mathrm{PD}_m(x)) - \varrho_{nm}^{\bar{X}} \cdot \Phi^{-1}(\mathrm{PD}_n(x))}{\sqrt{1-(\varrho_{nm}^{\bar{X}})^2}}\right) \cdot \mathrm{PD}_n'(x)$$

which implies the second statement of the theorem. □

The second term of the right-hand side of equation (10.12) captures the effect of a granular portfolio, i.e. it accounts for the difference between a finite and an infinite number of loans in the portfolio. This term represents the granularity adjustment and hence, it vanishes if the number N of loans in the portfolio tends to infinity provided that $\sum_{n=1}^{N} s_n^2 \to 0$ almost surely as $N \to \infty$. [109] prove the following assertion.

Theorem 10.1.4 *The conditional variance term $\sigma_{GA}^2(x)$ is given by*

$$\sigma_{GA}^2(x) = \sum_{n=1}^{N} s_n^2 \cdot \left[(\text{ELGD}_n^2 + \text{VLGD}_n^2) \cdot \text{PD}_n(x) \right. \tag{10.15}$$
$$\left. - \text{ELGD}_n^2 \cdot \Phi_2\left(\Phi^{-1}(\text{PD}_n(x)), \Phi^{-1}(\text{PD}_n(x)), \varrho_{nn}^{\bar{X}} \right) \right],$$

where $\Phi_2(.,.,.)$ is a bivariate normal distribution and $\varrho_{nm}^{\bar{X}}$ is the asset correlation between asset n and asset m conditional on $\bar{X}$. Its derivative equals

$$\frac{d}{dx}\sigma_{GA}^2(x) = \sum_{n=1}^{N} s_n^2 \cdot \text{PD}_n'(x) \cdot \left[(\text{ELGD}_n^2 + \text{VLGD}_n^2) \right.$$
$$\left. - 2 \cdot \text{ELGD}_n^2 \cdot \Phi\left(\frac{\Phi^{-1}(\text{PD}_n(x)) - \varrho_{nn}^{\bar{X}} \cdot \Phi^{-1}(\text{PD}_n(x))}{\sqrt{1 - (\varrho_{nn}^{\bar{X}})^2}} \right) \right]. \tag{10.16}$$

Proof. Recall that

$$L = \sum_{n=1}^{N} s_n \cdot \text{LGD}_n \cdot \mathbf{1}_{\{r_n \leq \Phi^{-1}(\text{PD}_n)\}}$$

and define the individual loss variables by

$$L_n = \text{LGD}_n \cdot \mathbf{1}_{\{r_n \leq \Phi^{-1}(\text{PD}_n)\}}.$$

Since LGD_n are independent between themselves as well as from all the other variables, all cross terms are zero and we can compute

$$\mathbb{V}[s_n L_n | \{X_1, \ldots, X_K\}] = \mathbb{V}[s_n \cdot \text{LGD}_n \cdot \mathbf{1}_{\{r_n \leq \Phi^{-1}(\text{PD}_n)\}} | \{X_1, \ldots, X_K\}]$$
$$= s_n^2 \cdot \mathbb{E}\left[\text{LGD}_n^2 | \{X_1, \ldots, X_K\} \right] \cdot \mathbb{E}\left[\mathbf{1}_{\{r_n \leq \Phi^{-1}(\text{PD}_n)\}} | \{X_1, \ldots, X_K\} \right]$$
$$- s_n^2 \cdot \mathbb{E}[\text{LGD}_n | \{X_1, \ldots, X_K\}]^2 \cdot \mathbb{E}\left[\mathbf{1}_{\{r_n \leq \Phi^{-1}(\text{PD}_n)\}} | \{X_1, \ldots, X_K\} \right]^2.$$

It follows that

10.1 Analytical Approximation Models

$$\mathbb{E}\left[\mathbb{V}[s_n L_n | \{X_1, \ldots, X_K\}] | \bar{X}\right]$$

$$= s_n^2 \cdot (\text{ELGD}_n^2 + \text{VLGD}_n^2) \cdot \mathbb{E}\left[\mathbf{1}_{\{r_n \leq \Phi^{-1}(\text{PD}_n)\}} | \bar{X}\right]$$

$$- s_n^2 \cdot \text{ELGD}_n^2 \cdot \mathbb{E}\left[\mathbf{1}_{\{r_n \leq \Phi^{-1}(\text{PD}_n)\}} \cdot \mathbf{1}_{\{r_n \leq \Phi^{-1}(\text{PD}_n)\}} | \bar{X}\right]$$

$$= s_n^2 \cdot (\text{ELGD}_n^2 + \text{VLGD}_n^2) \cdot \mathbb{E}\left[\mathbf{1}_{\{r_n \leq \Phi^{-1}(\text{PD}_n)\}} | \bar{X}\right]$$

$$- s_n^2 \cdot \text{ELGD}_n^2 \cdot \Phi_2\left(\Phi^{-1}(\text{PD}_n(x)), \Phi^{-1}(\text{PD}_n(x)), \varrho_{nn}^{\bar{X}}\right)$$

$$= s_n^2 \cdot (\text{ELGD}_n^2 + \text{VLGD}_n^2) \cdot \text{PD}_n(x)$$

$$- s_n^2 \cdot \text{ELGD}_n^2 \cdot \Phi_2\left(\Phi^{-1}(\text{PD}_n(x)), \Phi^{-1}(\text{PD}_n(x)), \varrho_{nn}^{\bar{X}}\right)$$

which proves equation (10.15). Equation (10.16) can be derived analogously to equation (10.14). □

Since the correction term, equation (10.11), is linear in the conditional variance $\sigma^2(x) = \sigma_\infty^2(x) + \sigma_{GA}^2(x)$ and its first derivative, the adjustment can be split into the sum of the *multi-factor adjustment* and the *granularity adjustment* term

$$\Delta \alpha_q = \Delta \alpha_q^\infty + \Delta \alpha_q^{GA},$$

where each term is obtained by inserting the corresponding conditional variance and its derivative in equation (10.11).[7] As we saw in Chapter 9, the granularity adjustment term $\Delta \alpha_q^{GA}$ vanishes as the number N of loans in the portfolio tends to infinity. Thus,

$$\alpha_q(\bar{L}) + \Delta \alpha_q^\infty$$

serves as an approximation for the quantile of L^∞.

10.1.2 Analytical Approximation for Expected Shortfall

[109] also applied their methodology to a different risk measure, namely Expected Shortfall (ES). As introduced in Section 2.4, ES is closely related to VaR where, instead of using a fixed confidence level as in VaR, one averages VaR over all confidence levels, which are larger than q. One major advantage of ES is that it gives information about the severity of losses, which occur with a probability less than $1 - q$. Recall that ES at a confidence level q for L is defined as the expected loss above the qth percentile.

$$\text{ES}_q(L) = \mathbb{E}[L|L \geq \alpha_q(L)] = \frac{1}{1-q} \int_q^1 \alpha_s(L) ds. \qquad (10.17)$$

---

[7] See [109], page 88.

In the last subsection we saw that $\alpha_q(L)$ can be approximated by the sum of $\alpha_q(\bar{L})$ and $\Delta\alpha_q(L)$. The analytical approximation can be extended to the case of ES. Therefore, we can decompose $\text{ES}_q(L)$ into ES for a comparable one-factor model $\text{ES}_q(\bar{L})$ and the ES multi-factor adjustment $\Delta\text{ES}_q(\bar{L})$. By exploiting the formula for the limiting loss distribution in the one-factor framework, [109] obtain[8]

$$\text{ES}_q(\bar{L}) = \mathbb{E}[\mu(\bar{X})|\bar{X} \leq \Phi^{-1}(1-q)] = \frac{1}{1-q}\int_{-\infty}^{\Phi^{-1}(1-q)} \phi(y)\mu(y)dy. \quad (10.18)$$

Again, we took advantage of the fact that $\mu(\bar{X}) = \bar{L}$ is a monotonic and deterministic function of $\bar{X}$.

In correspondence to equation (10.9), the multi-factor adjustment in the ES framework can be written as

$$\Delta\text{ES}_q(\bar{L}) = -\frac{1}{2(1-q)}\int_q^1 \frac{1}{\phi(x)} \cdot \frac{d}{dx}\left(\phi(x) \cdot \frac{\sigma^2(x)}{\mu'(x)}\right)\bigg|_{x=\Phi^{-1}(1-s)} ds. \quad (10.19)$$

All terms are defined in the same way as in the VaR framework. As with the multi-factor adjustment to VaR, the systematic and idiosyncratic parts can be separated by decomposing $\sigma^2(x)$, because ES is linear in $\sigma^2(x)$. This leads to an adjustment to ES of the form

$$\Delta\text{ES}_q(L) = \Delta\text{ES}_q^\infty(L) + \Delta\text{ES}_q^{GA}(L).$$

10.1.3 Performance Testing

[109] use several setups to compare the performance of the multi-factor adjustment with Monte Carlo simulations. The following discussion focuses on the main results of the performance testing without explaining the construction of the results in detail. The results base on a 99.9% quantile in the VaR framework.

The starting point is the multi-factor framework. The loans of a portfolio are grouped into different buckets. Within a bucket all characteristics (e.g. the factor loading, the weight ω of the global factor, the default probability, the mean and the standard deviation of loss given default as well as the composite factors Y_n) are equal across borrowers. The composite factor Y_n of borrower n is driven by one global systematic factor X_K as well as by its industry-specific factors $X_{k(n)}$, i.e.

$$Y_n = \omega \cdot X_K + \sqrt{1-\omega^2} \cdot X_{k(n)},$$

where $k(n)$ denotes the industry that borrower n belongs to. There exist $K-1$ industry-specific factors.

In the first setup it is assumed that the number of loans tends to infinity ($N \to \infty$). Therefore, this setup concentrates on the systematic part of the

---
[8] Implicitly, it is assumed that b_k and ω_n are the same for all s above q.

multi-factor adjustment since the idiosyncratic risk is diversified away. Hence, it only considers the difference between the multi-factor and the one-factor limiting loss distribution. Furthermore, we assume that the characteristics of all buckets are identical. We refer to this setup as the *homogenous case*. [109] define the *accuracy* as the ratio of the 99.9% quantile derived by the analytical approximation $\alpha_q(\bar{L}) + \Delta\alpha_q^\infty$ to the same quantile of L^∞ computed by Monte Carlo simulation. Their results show that accuracy quickly improves as the correlation ϱ between the composite risk factors increases. This observation is very intuitive since in the limit ($\varrho = 1$) the multi-factor model reduces to a one-factor model framework. Another interesting observation is that, at any fixed ϱ, the accuracy of the approximation improves as the number of factors increases. This observation can be justified by the fact that, if the number $K - 1$ of industry-specific factors increases to infinity, the composite factor is completely driven by the global systematic factor. Thus, the multi-factor model is equivalent to the single-factor model in this extreme case. The accuracy of the approximation is, for example, about 0.9 for $\varrho \leq 0.1$ and $N \leq 10$ and rises fast for increasing N or ϱ.

In the second setup it is assumed that the portfolio is divided into ten different buckets and the number of loans still tends to infinity. The most conspicuous distinction is that the multi-factor adjustment performs well even for very low levels of ϱ. Therefore, the accuracy is closer to one, compared to the case of homogeneous buckets.

In the last setup, [109] assume a finite number of loans and ten different buckets. Obviously, the approximation is better for a higher number of loans because the granularity adjustment becomes smaller in this case. Moreover, a high dispersion of the number of loans among the different buckets leads to larger inaccuracies between the multi-factor adjustment and the Monte Carlo simulation. A low dispersion implies that the idiosyncratic risk is almost diversified away. Hence, a high dispersion amplifies the presence of idiosyncratic risk, which implies a worse approximation.

10.1.4 Summary and Discussion

For a long time analytical methods for calculating portfolio VaR and ES have been limited to one-factor models. [109] extend these techniques for calculating VaR and ES to the multi-factor framework. Based on the analytical approximation method introduced in the granularity adjustment approach for a single-factor model, [109] develop a multi-factor adjustment for VaR and ES. In contrast to Monte Carlo simulations, the main advantage of such a multi-factor approximation to VaR is that portfolio VaR can be computed almost instantaneously. Obviously this can be of great advantage among risk management practitioners. Furthermore, the performance testing reveals that the multi-factor adjustment leads to excellent approximations throughout a wide range of model parameters. The differences in most setups between Monte Carlo simulation and the multi-factor approximation are very small.

Although the multi-factor adjustment closes a gap in the literature, it involves some questionable assumptions, which have to be discussed. First, it is assumed that the loss given default LGD_n is independent from all the other variables. However, there is strong empirical evidence that default rates and recovery rates $(1 - \mathrm{LGD}_n)$ are negatively correlated. For example, [64] used data from Moody's Default Risk Service Database for 1982-1997. Their results show a strong negative correlation between default rates and recovery rates for corporate bonds. According to [64] the intuition behind the correlation is as follows. If a borrower defaults on a loan, a bank's recovery may depend on the value of the collateral for the loan. But the value of collateral for loans usually depends on the state of the economy. If the economy experiences a recession, the recovery rate decreases while the default rate as well as the probability of default increase. Therefore, the assumption of independence between loss given default and the probability of default is questionable. By constructing a correlation between either variables, this approach could be extended.

Furthermore, asset returns are assumed to be standard normally distributed. This assumption simplifies the model and makes the calculations tractable. However, empirical investigations, for example by [59], reject the normal distribution because it is unable to model dependencies between extreme events. Thus, other multivariate distributions have been considered, for example the Student t-distribution and the normal inverse Gaussian distribution. It would be interesting to reformulate this model in a CreditRisk$^+$ setting to see the similarities not only to the GA technique but also to the saddle-point approximation method of Section 9.3.

10.2 Diversification Factor Models

In this section we discuss a semi-analytic model by [28] for measuring sector concentration. The starting point of their approach is the well known shortcoming of the ASRF model to neglect diversification effects on both obligor and sector level. [28] provide an extension of the ASRF model to a general multi-factor setting which can recognize diversification effects. They derive an adjustment to the single risk factor model in form of a scaling factor to the economic capital required by the ASRF model. This so-called *capital diversification factor* is a function depending on sector size and sector correlations. Sector size is reflected by an index similar to the Herfindahl-Hirschmann-Index, but capturing the relative size of the sector exposures and their credit characteristics. The diversification factor is estimated numerically using Monte Carlo simulations. Scaling the economic capital of the ASRF model by the es-

10.2.1 The Multi-Sector Framework

[28] measure sector concentration in loan portfolios based on a simplified version of the multi-factor Merton model presented in Section 3.2. We will refer to this simplified setting as to the *multi-sector framework*. The modifications will be explained in this subsection while we stick to the general notation of Section 3.2.

Each of the K factors in the model is identified with an industry sector and each borrower in the portfolio is assigned to one of these sectors. In the multi-factor Merton model, borrower's defaults exhibit a certain dependence structure due to sector-dependent systematic risk factors that are usually correlated. The asset returns r_n are modeled by equation (3.6). We simplify this framework by assuming that all borrowers in the same sector k are influenced by the same systematic risk factor Y_k. In other words, while in Chapter 3 every borrower n was influenced by its own composite factor Y_n, we now assume that, if two borrowers n and m belong to the same sector $k = k(n) = k(m)$, they are influenced by the same systematic risk factor Y_k. Hence the asset return r_n of borrower n can be expressed as

$$r_n = \beta_{k(n)} \cdot Y_{k(n)} + \sqrt{1 - \beta_{k(n)}^2} \cdot \varepsilon_n, \qquad (10.20)$$

where $k(n)$ denotes the sector to which borrower n belongs. Y_k is a standard normal random variable and the idiosyncratic shock ε_n is also standard normal. Hence, each sector is driven by a different risk factor. The overall portfolio loss is given by[9]

$$L = \sum_{k=1}^{K} \sum_{n \in \text{Sector } k} \text{EAD}_n \cdot \text{LGD}_n \cdot \mathbf{1}_{\{r_n \leq \Phi^{-1}(\text{PD}_n)\}}. \qquad (10.21)$$

Similarly to the model of [109] we now assume that the sector risk factors Y_k are correlated through a single macro risk factor $\bar{X}$[10]

$$Y_k = \varrho_k \cdot \bar{X} + \sqrt{1 - \varrho_k^2} \cdot \eta_k, \quad k = 1, \ldots, K, \qquad (10.22)$$

where η_k are standard normal variables independent of $\bar{X}$ and among each other. Each sector has a different correlation level ϱ_k with the macro risk factor. We refer to the ϱ_k as to the *inter- (or cross-) sector correlations* and to the β_k as to the *intra-sector correlations*.

---
[9] Compare equation (3.9).
[10] Compare equation (10.3).

Remark 10.2.1 Due to equations (10.20) and (10.22), the correlation between the asset returns r_n and r_m of obligors n and m belonging to sectors k and l, respectively, can be computed as

$$\text{Corr}(r_n, r_m) = \begin{cases} \beta_k^2 & k = l, \\ \beta_k \beta_l \varrho_k \varrho_l & k \neq l. \end{cases} \qquad (10.23)$$

Therefore, in order to derive the correlation between the asset returns in case of the same sector, only an estimate of the intra-sector correlation β_k is needed where else also the cross-sector correlation parameters ϱ_k need to be estimated.

Proof of Equation (10.23). The covariance of r_n and r_m for firms n and m belonging to sectors k and l, respectively, can be computed as[11]

$$\text{Cov}(r_n, r_m) = \mathbb{E}[r_n r_m] - \mathbb{E}[r_n] \cdot \mathbb{E}[r_m]$$

$$= \beta_k \beta_l \cdot \mathbb{E}[Y_k Y_l] - \beta_k \beta_l \cdot \mathbb{E}[Y_k] \cdot \mathbb{E}[Y_l]$$

$$= \beta_k \beta_l \cdot \varrho_k \varrho_l \cdot \mathbb{E}[\bar{X}^2] + \beta_k \beta_l \cdot \sqrt{1 - \varrho_k^2} \cdot \sqrt{1 - \varrho_l^2} \cdot \mathbb{E}[\eta_k \eta_l]$$

$$= \beta_k \beta_l \cdot \varrho_k \varrho_l + \beta_k \beta_l \cdot \sqrt{1 - \varrho_k^2} \cdot \sqrt{1 - \varrho_l^2} \cdot \delta_{kl}$$

$$= \begin{cases} \beta_k^2 & \text{for } k = l \\ \beta_k \beta_l \varrho_k \varrho_l & \text{for } k \neq l \end{cases}$$

where we used the independence and standard normality of all factors. Noting that the variance $\mathbb{V}[r_n] = 1$ for all n, we have proved the assertion for the correlation between r_n and r_m. $\square$

From equation (10.5) we obtain the conditional expectation of the portfolio loss in this model. Due to Theorem 4.1.6 we have for asymptotically fine-grained sector portfolios that the stand-alone q^{th} quantile of the portfolio loss for a given sector k equals

$$\alpha_q(L) = \sum_{n \in \text{Sector } k} \text{EAD}_n \cdot \text{ELGD}_n \cdot \Phi\left(\frac{\Phi^{-1}(\text{PD}_n) + \beta_k \Phi^{-1}(q)}{\sqrt{1 - \beta_k^2}}\right).$$

Note that here ϱ_k is neglected as we only consider the sector portfolio and its stand-alone VaR. Hence, under the assumptions of the ASRF model the economic capital for sector k takes the same form as in the IRB approach, i.e.

$$EC_k(q) = \sum_{n \in \text{Sector } k} \text{EAD}_n \cdot \text{ELGD}_n \cdot \left[\Phi\left(\frac{\Phi^{-1}(\text{PD}_n) + \beta_k \Phi^{-1}(q)}{\sqrt{1 - \beta_k^2}}\right) - \text{PD}_n\right]. \qquad (10.24)$$

---

[11] δ_{kl} denotes the Kronecker symbol which is equal to 1 for $k = l$ and 0 otherwise.

10.2.2 The Capital Diversification Factor

For simplicity [28] assume an average factor correlation for all sectors, i.e. $\varrho_k = \varrho$ for all $k = 1, \ldots, K$. Under the assumption of perfect correlation between all sectors, i.e. $\varrho = 1$, which is equivalent to the ASRF model setting, the economic capital for the whole portfolio is the sum of the economic capital of the individual sectors, i.e.

$$\mathrm{EC}^{\mathrm{sf}} = \sum_{k=1}^{K} \mathrm{EC}_k, \qquad (10.25)$$

where the superscript *sf* indicates the single risk factor setting. The following definition is due to [28].

Definition 10.2.2 (Capital Diversification Factor)
The *capital diversification factor* DF is defined as the ratio of the economic capital computed using the multi-factor setting and the stand-alone capital

$$\mathrm{DF} = \frac{\mathrm{EC}^{\mathrm{mf}}}{\mathrm{EC}^{\mathrm{sf}}}, \qquad (10.26)$$

with $0 < DF \leq 1$.

As the closed form expression for DF can be quite intricate, the aim of [28] is to derive an approximation for DF by a scalar function of a small number of parameters which then need to be calibrated. [28] presume that diversification results mainly from two sources. The first source is the relative size of various sector portfolios and the second source lies in the correlation ϱ between sectors. To measure the relative size of the sector portfolios, [28] use the following index.

Definition 10.2.3 (Capital Diversification Index)
The *capital diversification index* CDI is the sum of squares of the economic capital weights in each sector

$$\mathrm{CDI} = \frac{\sum_{k=1}^{K} \mathrm{EC}_k^2}{(\mathrm{EC}^{\mathrm{sf}})^2} = \sum_{k=1}^{K} \omega_k^2, \qquad (10.27)$$

where $\omega_k = \mathrm{EC}_k / \mathrm{EC}^{\mathrm{sf}}$ is the weight of sector k with respect to the one-factor capital.

This is simply the Herfindahl-Hirschman concentration index[12] applied to the stand-alone economic capital of each sector, while the correlation ϱ between sectors is not taken into account. The CDI considers both the relative size of the sector exposure and their credit characteristics.

---

[12] See Chapter 8 for details.

Concerning the cross-sector correlation, [28] show that, for correlated sectors with uniform correlation parameter ϱ, the standard deviation σ_P of the portfolio loss distribution can be expressed as

$$\sigma_P = \sqrt{(1-\varrho^2)\cdot \mathrm{CDI} + \varrho^2} \cdot \sum_{k=1}^{K} \sigma_k,$$

where σ_k denotes the volatility of the credit loss for sector k. For normally distributed losses a similar relation holds for the economic capital

$$\mathrm{EC}^{\mathrm{mf}} = \mathrm{DF}^N(\mathrm{CDI},\varrho) \cdot \mathrm{EC}^{\mathrm{sf}},$$

where $\mathrm{DF}^N = \sqrt{(1-\varrho^2)\cdot \mathrm{CDI} + \varrho^2}$ is the diversification factor for normally distributed losses. Motivated by this relationship, [28] approximate the diversified economic capital for a given confidence level q as some factor DF, depending on CDI and ϱ, multiplied with $\mathrm{EC}^{\mathrm{sf}}$, i.e.

$$\mathrm{EC}^{\mathrm{mf}}(q) \approx \mathrm{DF}(\mathrm{CDI},\varrho) \cdot \mathrm{EC}^{\mathrm{sf}}(q). \tag{10.28}$$

If the sectors are perfectly correlated, i.e. $\varrho = 1$, then the model reduces to the single-factor setting, leading to the constraints

$$\mathrm{DF}(\mathrm{CDI},1) = 1 \quad \text{and} \quad \mathrm{DF}(1,\varrho) = 1. \tag{10.29}$$

If the diversification adjustment is allocated uniformly across sectors, the economic capital contribution for a given sector is $\mathrm{DF} \cdot \mathrm{EC}_k$.

10.2.3 Marginal Capital Contributions

As mentioned in the beginning of Subsection 10.2.2, in the single-factor setting the capital contribution to a given sector equals its stand-alone capital EC_k and total capital is just the sum of the stand-alone capitals. This, however, does not hold in the multi-factor framework where capital contributions need to be calculated on a marginal basis. Analogously to equation (10.25), [28] show that the multi-factor adjusted capital is a linear combination of the stand-alone economic capital of each sector. That is the multi-factor adjusted economic capital can be decomposed as

$$\mathrm{EC}^{\mathrm{mf}} = \sum_{k=1}^{K} \mathrm{DF}_k \cdot \mathrm{EC}_k,$$

where

$$\mathrm{DF}_k = \frac{\partial \, \mathrm{EC}^{\mathrm{mf}}}{\partial \, \mathrm{EC}_k}, \quad k = 1,\ldots K,$$

is the so-called *marginal sector diversification factor*. It is obtained by an application of Euler's Theorem.[13] A closed-form expression for DF_k can be derived by an application of the chain rule as

$$\mathrm{DF}_k = \mathrm{DF} + 2\frac{\partial \mathrm{DF}}{\partial \mathrm{CDI}} \cdot \left(\frac{\mathrm{EC}_k}{\mathrm{EC}^{\mathrm{sf}}} - \mathrm{CDI}\right) + 2\frac{\partial \mathrm{DF}}{\partial \bar{\varrho}} \cdot \frac{1 - \frac{\mathrm{EC}_k}{\mathrm{EC}^{\mathrm{sf}}}}{1 - \mathrm{CDI}} \cdot \left(\bar{Q}_k - \bar{\varrho}\right). \quad (10.30)$$

Here $\bar{\varrho}$ is the average cross-sector correlation. The correlation between sectors is represented by the $K \times K$ matrix

$$\begin{pmatrix} 1 & q_{12} & \cdots & q_{1K} \\ q_{21} & 1 & \cdots & q_{2K} \\ \vdots & \vdots & & \vdots \\ q_{K1} & q_{K2} & \cdots & 1 \end{pmatrix}$$

with q_{kl} denoting the correlation between sectors k and l for $k, l = 1, \ldots, K$. [28] define the average cross-sector correlation as

$$\bar{\varrho} = \frac{\sum_{k=1}^{K} \sum_{l \neq k} q_{kl} \cdot \mathrm{EC}_k \cdot \mathrm{EC}_l}{\sum_{k=1}^{K} \sum_{l \neq k} \mathrm{EC}_k \cdot \mathrm{EC}_l} = \frac{\sum_{k=1}^{K} \sum_{l \neq k} q_{kl} \cdot \omega_k \omega_l}{\sum_{k=1}^{K} \sum_{l \neq k} \omega_k \omega_l}. \quad (10.31)$$

Thus

$$\bar{Q}_k = \frac{\sum_{l \neq k} q_{kl} \cdot \mathrm{EC}_l}{\sum_{l \neq k} \mathrm{EC}_l}$$

represents the average correlation of sector factor k to the rest of the systematic sector factors in the portfolio.

Formula (10.30) shows that the diversification effect consists of three components; the overall portfolio diversification DF, the sector size and the sector correlation,

$$\mathrm{DF}_k = \mathrm{DF} + \Delta\mathrm{DF}_{size} + \Delta\mathrm{DF}_{corr}. \quad (10.32)$$

In the special case where all sectors have the same constant cross-sector correlation coefficient ϱ, we have

$$\mathrm{DF}_k = \mathrm{DF} + 2 \cdot \frac{\partial \mathrm{DF}}{\partial \mathrm{CDI}} \cdot \left(\frac{\mathrm{EC}_k}{\mathrm{EC}^{\mathrm{sf}}} - \mathrm{CDI}\right), \quad (10.33)$$

which is only determined by the overall effect DF and the size effect $\Delta\mathrm{DF}_{size}$. Moreover, the formula tells us that sectors with small stand-alone capital and lower correlation than the average, i.e. $\mathrm{EC}_k / \mathrm{EC}^{\mathrm{sf}} < \mathrm{CDI}$ and $\bar{Q}_k < \bar{\varrho}$, will result in $\mathrm{DF}_k < \mathrm{DF}$, which means that these sectors will have a lower capital requirement than in the case where the diversification effect is allocated evenly across sectors.

---

[13] Note that we can apply Euler's Theorem here, since the multi-factor capital formula (10.28) is a homogeneous function of degree one.

10.2.4 Parameterization

The parameterization is based on Monte-Carlo simulations. Recall that the sector portfolios are assumed to be infinitely fine-grained. Furthermore, [28] assume for the simulations that, within each sector k, all obligors have the same default probability PD_k, the same exposure at default EAD_k and the same intra-sector correlation β_k. Loss given defaults are fixed to ELGD = 100%. The total portfolio exposure is normalized to one, $\sum_{k=1}^{K} \mathrm{EAD}_k = 1$.

For simplicity, [28][14] assume that all sectors have the same correlation level with the macro risk factor $\bar{X}$, which means that in equation (10.22) we have $\varrho_k = \varrho$ for all $k = 1, \ldots, K$. The parameterization is based on the following steps.

1. In each simulation, sample independently the sector exposure sizes EAD_k and the default probabilities PD_k for $k = 1, \ldots, K$. The intra-sector correlations β_k are derived according to the Basel II formula for the wholesale exposures.
2. For each portfolio the sector stand-alone capitals EC_k and the sum of them, $\mathrm{EC}^{\mathrm{sf}} = \sum_{k=1}^{K} \mathrm{EC}_k$, and the CDI are computed analytically according to the sampled parameters, while the multi-factor adjusted economic capital $\mathrm{EC}^{\mathrm{mf}}$ is derived from Monte Carlo simulations.
3. Repeat the simulations by varying the number K of sectors and the cross-sector correlation $\varrho \in [0, 1]$.
4. Based on the simulated data, estimate the diversification effect DF as a function of CDI and ϱ.

According to the constraints (10.29) for the diversification factor, [28] presume that the functional form of $\mathrm{DF}(\mathrm{CDI}, \varrho)$ is a polynomial of the form

$$\mathrm{DF}(\mathrm{CDI}, \varrho) = 1 + \sum_{i,j \geq 1} a_{i,j} \cdot (1 - \varrho)^i \cdot (1 - \mathrm{CDI})^j \quad (10.34)$$

with coefficients $a_{i,j}$. Based on 22,000 simulated portfolios they derive the following parameter estimations

$$\mathrm{DF}(\mathrm{CDI}, \varrho) = 1 + \sum_{i,j=1}^{2} a_{i,j} \cdot (1 - \varrho)^i \cdot (1 - \mathrm{CDI})^j$$

with $a_{1,1} = -0.852$, $a_{1,2} = 0$, $a_{2,1} = 0.426$, $a_{2,2} - 0.481$.

[28] further conduct parameterization tests for portfolios with different sector correlation levels. In contrast to the first parameterization test, the ϱ_k are simulated as independent uniform random variables in the interval $[0, 1]$. According to the authors they get the same results with good performance.

---

[14] The description of the methodology in this section is based on the presentation of Herrero on Oct. 19, 2006 by Risklab-Madrid.

The diversification effect depends on the CDI and the average cross-sector correlation $\bar{\varrho}$, and $\mathrm{DF}(\mathrm{CDI}, \bar{\varrho})$ can be approximated with the same polynomial function[15]

$$\mathrm{DF}(\mathrm{CDI}, \bar{\varrho}) = 1 + \sum_{i,j=1}^{2} a_{i,j} \cdot (1 - \varrho)^i \cdot (1 - \mathrm{CDI})^j. \qquad (10.35)$$

Together with equation (10.30) the marginal sector diversification factors can be calculated.

10.2.5 Application to a Bank Internal Multi-Factor Model

[28] show that their model can be fitted to a bank internal multi-factor model by calculating the implied parameters like intra-sector correlations β_k, the average cross-sector correlation $\bar{\varrho}$ and the relative sector size and cross-sector correlations. They give an example with four sectors, S1-S4. The corresponding parameters are summarized in Table 10.1. The default probability in the first two sectors equals 1% while the last two sectors have a PD of 0.5%. ELGD is assumed to be 100%. The exposures of each sector are given in the second column of Table 10.1 and the sector stand-alone (SA) economic capitals calculated by the bank internal model are listed in the third column. With these data the implied inter-sector correlations Q_k can be derived and are listed in the last column. The fourth sector, for example, accounts for one tenth of total portfolio exposure while it has only 4% share of stand-alone capital. This effect can be explained by its low PD and also by its low implied intra-sector correlation β_k, displayed in column four. All other sectors have a much higher share of stand-alone capital which is due to the fact that their implied intra-sector correlations β_k are all higher. In line with these observations is the fact, that the capital diversification index CDI has a value close to one third, implying that there are roughly three effective sectors.

Table 10.1. Multi-factor economic capital and implied correlations (See [27], Table 10.)

Sector	EAD	SA Capital (One-Factor)	Implied β_k	Capital % (Flat Beta=54.6%)	Economic Capital %	Implied Q_k
S1	25	35.3%	20.1%	36.1%	31.9%	37.1%
S2	25	21.5%	12.4%	19.0%	17.2%	42.8%
S3	40	39.6%	21.9%	42.3%	47.5%	74.2%
S4	10	3.7%	8.6%	2.6%	3.4%	89.0%
Total	100					
		SA Capital	CDI	DF	Capital	Implied $\bar{\varrho}$
		9.7	32.9%	75.5%	7.3%	54.9%

---

[15] The only difference is that ϱ is replaced by $\bar{\varrho}$.

The EC^{mf} is calculated using the bank internal model and amounts to 7.3% of total exposure. DF is then calculated as $\text{EC}^{\text{mf}}/\text{EC}^{\text{sf}}$ which is equal to 75.5%. The implied average correlation $\bar{\varrho}$ is then derived from DF by linear interpolation.[16] The relative economic capital of each sector, calculated by the bank internal model under the assumption that all sector correlations are equal to the average of 54.9%, is listed in column 6. The parameters $\bar{Q}_k$ in equation (10.30) are listed in the last column.

Table 10.2. Marginal diversification factors for flat inter-sector correlation (See [27], Table 11.)

Sector	DF_k	Portfolio Diversification	Sector Size	Sector Correlation
S1	77.5%	75.6%	1.8%	0%
S2	66.9%	75.6%	-8.7%	0%
S3	80.8%	75.6%	5.2%	0%
S4	53.3%	75.6%	-22.3%	0%

Tables 10.2 and 10.3 show the decomposition of the sector diversification factor for flat inter-sector correlations and implied inter-sector correlations, respectively. In comparison to the stand-alone case, the size component of the sector diversification factor increases contributions for the largest sectors S1 and S3 while it decreases them for the smaller sectors S2 and S4. Compared to these results, Table 10.3 implies that using implied inter-sector correlations decreases the marginal sector diversification factors for sectors S1 and S2 while it increases DF_k for S3 and S4. This is due to a negative, respectively positive, sector correlation diversification component, reported in the last column.

Table 10.3. Marginal diversification factors for implied inter-sector correlations (See [27], Table 11.)

Sector	DF_k	Portfolio Diversification	Sector Size	Sector Correlation
S1	68.4%	75.6%	1.8%	-9.1%
S2	60.7%	75.6%	-8.7%	-6.2%
S3	90.6%	75.6%	5.2%	9.8%
S4	70.7%	75.6%	-22.3%	17.4%

---

[16] See [27], p. 30.

10.2.6 Discussion

The model of [28] provides an approach to compute an analytical multi-factor adjustment for the ASRF model which requires a very small number of parameters. Compared to the method of [109], introduced in Section 10.1, their approach might be less general and it requires some calibration work, however, with the benefit of being much simpler and leading to less complicated expressions for capital contributions.

Once the diversification factor is calibrated the calculations are fast and simple which is invaluable for real-time capital allocation. Concentration risk is determined by the "effective sector size" CDI and average cross sector correlation $\bar{\varrho}$. The estimated parameters are derived from Monte Carlo simulations based on the assumptions of the multi-factor Merton model. The parameters $\bar{\varrho}$ and CDI are chosen as explaining factors for the diversification factor DF because both factors are presumed to be quite stable over time and might vary only to a very small extend. This conjecture, however, is not verified through empirical tests.

The diversification factor model proposed in [28] is based on the assumptions of the ASRF model. However, since realistic portfolios are rarely infinitely granular, ignoring name concentration can potentially lead to an underestimation of total portfolio credit risk. The model proposed by [28], however, does not account for this source of concentration risk and an extension of their method to non-granular sectors by increasing intra-sector correlations seems to be problematic.

Furthermore, the model does not allow for obligor specific asset correlations. In addition, under the simplifying assumption of average cross-sector correlations, the model cannot distinguish between a portfolio which is highly concentrated towards a sector with a high correlation with other sectors, and another portfolio which is equally high concentrated, but towards a sector which is only weakly correlated with other sectors. This, however, could change the amount of sector concentration risk in the portfolio considerably.

All in all the model provides a simple and intuitive way to compute and allocate EC. The main challenge, however, might be to remove reliance on calibration which would make the approach much more valuable.

11

Empirical Studies on Concentration Risk

Based on an empirical analysis of stock market indices, [28] investigate the possible range of asset correlations within and across developed and emerging economies. They perform a principal component analysis (PCA)[1] of the individual stock market index returns in developed and emerging economies to underpin the argument that there is rarely only a single risk factor which affects all obligors in the same way and to the same extent. [28] show that there are multiple factors that affect developed and emerging economies and that these factors are not the same in both economies. In particular, they show that a single factor accounts for only 77.5% of the variability of the developed markets, and three factors are required to explain more than 90% while for emerging markets the first factor accounts for only 47% of variability and seven factors are required to explain more than 90%.[2]

These results demonstrate that a more general framework than a simple one-factor setting has to be considered to account for sector concentration risk. In this chapter we present some empirical studies which demonstrate the effects of name and sector concentration on portfolio credit risk. We start in Section 11.1 with an analysis based on the multi-factor Merton model and the associated multi-factor adjustment technique of [109]; already presented in the former chapter. Their results show that sector concentration can dramatically increase economic capital which underlines the importance of diversification on sector level. The multi-factor adjustment approach of [109] proves to be quite reliable and stable compared to Monte Carlo based simulations for credit portfolio VaR based on multi-factor models.

In Section 11.2 we discuss the influence of systematic and idiosyncratic risk on the portfolio loss distribution based on an empirical study by [25]. The authors show that idiosyncratic risk has a minor impact on portfolio

---

[1] PCA is a statistical technique for finding patterns in data with high dimensions. The purpose of PCA is to reduce the dimensionality of data without much loss of information.

[2] See [27], p. 6-7.

VaR for large portfolios while it can meaningfully increase portfolio VaR for small portfolios. This is broadly in line with the already presented empirical results of Section 9.1. Moreover, their analysis supports the findings of [48] that sector concentration has a rather strong impact on the portfolio loss distribution. Moreover, they show that the contribution of systematic risk to portfolio VaR is almost independent of the portfolio size.

[25], furthermore, investigate the extend to which credit portfolio VaR can be explained by simple indices of name and sector concentration. They show that simple concentration indices are positively correlated with portfolio VaR even after controlling for differences in average credit quality and portfolio size.[3]

11.1 Sector Concentration and Economic Capital

Considering the findings above, [48] address two remaining questions in the context of sector concentration risk. First, they investigate to what extent sector concentration causes an increase in the economic capital associated with a bank's credit portfolio. Second, [48] analyze whether sector concentration risk can be measured by a model that does not require MC simulations.

In order to answer the first question, [48] simulate economic capital for a sequence of credit portfolios. Starting from a benchmark loan portfolio, which reflects the average sectoral distribution of loans in the German banking system, [48] construct six subsequent loan portfolios by stepwise increasing the concentration of the loan portfolios in one specific sector. This procedure is repeated until the last portfolio constitutes a single-sector portfolio, i.e. the sixth portfolio exhibits a total concentration in a particular sector. Since the average distribution of exposures is similar to the one in France, Belgium and Spain, the results are supposed to hold to a large extent for these countries as well. Notice that some of the generated portfolios reflect the degree of sector concentration of corporate loan portfolios of actual German banks. In their simulations, [48] find that economic capital indeed increases by nearly 37% compared with the benchmark portfolio and even more for the single-sector portfolio. The findings underline the importance to account for sector concentration when modeling credit risk.

Given the results outlined above, finding an analytical method for computing economic capital, which avoids the use of computationally burdensome Monte Carlo simulations, would be desirable. In this context, [48] examine the accuracy of the analytical approximation method of [109] for economic capital. [109] provide a tractable closed-form solution in a multi-factor framework, which at the same time allows to incorporate the effects of sector concentration (see Section 10.1 for details). In order to evaluate the accuracy of the economic capital approximation, [48] compare the economic capital values

---

[3] See [24], p. 1.

11.1 Sector Concentration and Economic Capital

from simulations with economic capital proxies computed according to [109]. Since this comparison shows that the economic capital approximation using the analytical model performs quite well, banks and supervisors are offered an approach to approximate economic capital by a rather simple formula and without computationally intensive Monte Carlo simulations.

11.1.1 The Model Framework

[48] measure sector concentration in loan portfolios based on the multi-factor adjustment method of [109] presented in Section 10.1. Within the multi-sector framework of Subsection 10.2.1, they further assume that each loan has the same relative portfolio weight $s_n = 1/N$ for all $n = 1, \ldots, N$. Denote by N_k the number of borrowers in sector k. Then the exposure share of sector k is given by N_k/N. The authors assume the expected loss severity and the probability of default to be equal for all borrowers in the same sector, i.e.

$$\text{ELGD}_{k(n)} = \text{ELGD}_n = \mathbb{E}\left[\text{LGD}_n\right] \quad \text{and} \quad \text{PD}_{k(n)} = \text{PD}_n\,.$$

Then the overall portfolio loss is given by[4]

$$L = \sum_{k=1}^{K} \sum_{n=1}^{N_k} s_n \cdot \text{LGD}_{k(n)} \cdot \mathbf{1}_{\left\{r_n \leq \Phi^{-1}(\text{PD}_{k(n)})\right\}}. \tag{11.1}$$

In order to obtain the loss distribution, [48] run a Monte Carlo simulation by generating asset returns from equation (10.20). If the asset returns cross the threshold $c_{k(n)} := \Phi^{-1}(\text{PD}_{k(n)})$, the according borrower defaults. The overall portfolio loss in each simulation run is calculated from equation (11.1). The portfolio loss distribution can then be used to calculate the economic capital (EC_{sim}) by subtracting the expected loss EL from the qth quantile $\alpha_q(L)$ of the loss distribution as

$$\text{EC}_{sim} = \alpha_q(L) - \text{EL}\,.$$

[48] compare the simulated results for economic capital (EC_{sim}) of various portfolios with the economic capital (EC_{MFA}) obtained via the multi-factor adjustment technique of [109], presented in Section 10.1, in order to examine the accuracy of the analytical approximation method.

11.1.2 Data Description and Portfolio Composition

[48] base their analysis on banks' loan portfolio data extracted from the European credit register. To analyze the sectoral concentration in realistic loan portfolios, all individual borrowers have to be assigned to a particular sector. [48] use a mapping between the *NACE classification scheme* and the *Global*

---
[4] Compare equation (3.9) or (10.21).

Industry Classification Standard (GICS). By the German Banking Act (*Kreditwesengesetz*) all German credit institutions are required to report quarterly exposure amounts (and national codes that are compatible with the NACE codes) of borrowers whose indebtedness to them amounts to 1.5 million Euro or more at any time during the three calender months preceding the reporting date.[5] Moreover, individual borrowers, who exhibit roughly the same risk, are aggregated into so-called *borrower units*. The borrower unit is assigned to that sector to which it has the dominating exposure. The Global Industry Classification Standard comprises ten sectors. [48] discard exposures to the "Financial" sector due to its special characteristics. Furthermore, they split the "Industrial" sector into three sub-sectors since this sector is rather heterogeneous. Hence, exposures are assigned to one of 11 different business sectors which are reported in Table 11.1.

As a benchmark portfolio [48] have aggregated exposures of loan portfolios of 2224 German banks in September 2004. Therefore, the resulting portfolio reflects the average sectoral concentration of the German banking system and can, henceforth, be considered as a well-diversified portfolio.[6] [48] focus solely on the impact of sectoral concentration on economic capital. Therefore, they assume that the benchmark portfolio is completely homogeneous across sectors, meaning that exposure sizes are all equal and a uniform default probability of PD = 2% and uniform loss given default of ELGD = 45% is assumed. Although the model in principle allows that each borrower is exposed to different sectors, [48] assume that each obligor can be assigned to a single sector.

In addition to the benchmark portfolio, a series of portfolios with increasing sector concentration is constructed. Starting from the benchmark portfolio, five subsequent portfolios are constructed by repeatedly reallocating one third of each sector exposure to the sector "Capital Goods". Portfolio 6 is created as a single-sector portfolio where all exposures are cumulated to the sector "Capital Goods". Table 11.1 shows the sector concentrations of the benchmark portfolio and the six subsequent portfolios with increasing sector concentration. The HHI of some of the stylized portfolios corresponds roughly to the observed ones in real-world portfolios. Portfolio 6 represents a completely concentrated portfolio where all exposures are assigned to a single sector. Note that these portfolios are also completely homogeneous across sectors as they have been constructed from the homogeneous benchmark portfolio.

Given that asset correlations are usually not observable, [48] estimate them using correlations of stock index returns. More precisely, inter-sector correlations are based on weekly returns of the MSCI EMU industry indices which

---

[5] See [48], p. 8.
[6] Since the sector concentration for Belgium, France and Spain are quite similar, the obtained results are supposed to be transferable to these countries.

11.1 Sector Concentration and Economic Capital 135

Table 11.1. Benchmark portfolio and sequence of portfolios with increasing sector concentration (See [48], Table 2)

	Benchmark PF	PF 1	PF 2	PF 3	PF 4	PF 5	PF 6
Energy	0%	0%	0%	0%	0%	0%	0%
Materials	6%	4%	3%	2%	2%	1%	0%
Capital Goods	12%	41%	56%	71%	78%	82%	100%
Commercial Services & Supplies	34%	22%	17%	11%	8%	7%	0%
Transportation	7%	5%	4%	2%	2%	1%	0%
Consumer Discretionary	15%	10%	7%	5%	4%	3%	0%
Consumer Staples	6%	4%	3%	2%	2%	1%	0%
Health Care	9%	6%	5%	3%	2%	2%	0%
Information Technology	3%	2%	2%	1%	1%	1%	0%
Telecommunication Services	1%	1%	1%	0%	0%	0%	0%
Utilities	7%	4%	3%	2%	2%	1%	0%
HHI	17.6	24.1	35.2	51.5	61.7	68.4	1

correspond to the 11 sectors. From these data the correlations between different borrowers can be calculated by multiplying the corresponding sector correlations with the factor weights. More precisely, for borrower n in sector k and borrower m in sector l the correlation is obtained by multiplying the correlation between sectors k and l by the factor weights β_n and β_m. Since the determination of the factor weights is much more difficult than the estimation of the sector correlations, [48] assume a unique factor weight $\beta = \beta_n = 0.50$ for all exposures. Hence the intra-sector asset correlation, i.e. the correlation of two obligors within the same sector, is fixed at 25%.

11.1.3 Impact of Sector Concentration on Economic Capital

The purpose of this subsection is to determine the impact of an increase in sector concentration on economic capital, computed as the difference between the 99.9% quantile of the loss distribution and the expected loss. EC is simulated by Monte Carlo simulation both for portfolios that comprise only corporate credits and for banks' total credit portfolios where the total portfolio includes 30% corporate credit and 70% other credit (to retail, sovereign, etc.). [48] compute the EC for the total portfolio as the sum of EC for the corporate portfolio and the EC for the remaining portfolio. The results for the benchmark portfolio and the more concentrated portfolios are presented in Table 11.2. Concerning the corporate portfolios, the results show that economic capital increases by 20% from the benchmark portfolio to Portfolio 2 and by even 35% for Portfolio 5. The second row of Table 11.2 demonstrates that an increase in sector concentration has less impact on the economic capital when considering the total loan portfolio.

In addition, [48] carry out several robustness checks, for instance by using a different factor correlation matrix or different default probabilities. The results of the robustness checks are shown in [48], Table 6. It turns out that the relative increase in economic capital is similar to the results above, even though the absolute values vary. These empirical results indicate the importance of taking sector concentration into account when calculating economic capital.

Table 11.2. Impact of sector concentration on EC_{sim} for the sequence of corporate and total credit portfolios (in %) (See [48], Table 4.)

	Benchmark	PF 1	PF 2	PF 3	PF 4	PF 5	PF 6
Corporate portfolio	7.8	8.8	9.5	10.1	10.3	10.7	11.7
Total portfolio	8.0	8.2	8.5	8.7	8.8	8.9	9.2

11.1.4 Robustness of EC Approximations

[48] compare the approximations for economic capital EC_{MFA}, computed by the multi-factor adjustment model of [109], with the economic capital EC_{sim} obtained from Monte Carlo simulation. As a starting point they assume a highly granular exposure distribution and homogeneity within each sector. The analytical approximation of EC was derived under exactly these assumptions. Therefore, the approximation should yield good estimates compared with the simulated results. To quantify the impact of sector concentration on EC, [48] compare the results for the approximated and the simulated EC with the EC computed in the ASRF model. That is they also compute EC^* as the 99.9% quantile of portfolio loss $\bar{L} = \mathbb{E}[L|\bar{X}]$ in the one-factor Merton model where $\bar{X}$ denotes the effective single systematic risk factor as in Section 10.1. The results of Table 11.3 show that EC^* and EC_{MFA} perform very well for approximating EC_{sim} for the benchmark portfolio and the portfolios with higher sector concentrations. This is supported by the fact that the relative error of EC_{MFA} is 1.3% at most.

[48] also investigate the robustness of the approximation with respect to the sector distribution, the factor correlations, the factor weights, the number of factors and the sector default probability. Under the assumption of homogeneous factor correlation matrices, [48] vary the entries of the matrix outside the main diagonal between 0 and 1. The extreme case with all correlations equal to 1 corresponds to a single factor model. The numerical values for the three EC calculations are not reported here in order to conserve space.[7] The results reveal that the approximations EC^* and EC_{MFA} are still extremely accurate proxies for EC_{sim}. The relative error of EC_{MFA} compared to EC_{sim} is at most 2.5%. In addition, an increase of factor correlations leads to a

---

[7] The values can be found in [48], Table 8.

11.1 Sector Concentration and Economic Capital

Table 11.3. Comparison of EC^*, EC_{MFA} and EC_{sim} (in percent of total exposure) (See [48], Table 7.)

Portfolio	EC^*	EC_{MFA}	EC_{sim}	Relative error of EC_{sim} and EC_{MFA}
Benchmark	7.8	7.9	7.8	1.30%
Portfolio 1	8.7	8.8	8.8	0.00%
Portfolio 2	9.4	9.4	9.5	-1.10%
Portfolio 3	10.1	10.1	10.1	0.00%
Portfolio 4	10.5	10.5	10.3	1.90%
Portfolio 5	10.7	10.7	10.7	0.00%
Portfolio 6	11.6	11.6	11.7	-0.90%

convergence of EC^* and EC_{MFA}. This is intuitive since increasing the factor correlations leads to a convergence of the multi-factor model against the single-factor model.

[48] also investigate the influence of the factor weight β and the number of sectors on the economic capital. They show that increasing the factor weights is associated with an increase in economic capital, but the accuracy of the EC approximations is not deteriorated. Moreover, given a certain number of sectors, economic capital increases with factor correlation. Their results indicate that for given factor correlations both EC^* and EC_{sim} decrease with increasing number of sectors. In line with the findings above, the approximation quality of EC^* and EC_{MFA} is confirmed, since relative errors are less than 2% and 1%, respectively.

Finally, the approximation quality is tested by assuming PD heterogeneity on a sector level. For this purpose, historical default rates are employed instead of unique default probabilities of 2%. For all EC calculations the results are very close to the ones in Table 11.3. Thus, the results obtained for homogeneous default probabilities should also hold for the case of heterogeneous sector-dependent default probabilities.

Of course the assumption of high granularity and uniform default probabilities in each sector is questionable in practice. [48] investigate the impact of lower granularity, restricting the analysis to the benchmark portfolio and Portfolio 6. For the examination two levels of granularity are considered. First, [48] construct a portfolio such that the distribution of exposures represents a typical small, regional German bank with average granularity. Second, in order to obtain an upper limit for the possible impact of granularity, a portfolio is created such that its exposure shares are the most concentrated possible

138   11 Empirical Studies on Concentration Risk

under the EU *large exposure rules*.[8] That is the contribution of every loan to regulatory capital must not exceed 25% while the sum of all those loans, which contribute more than 10% to regulatory capital, must not require more than 8 times the regulatory capital. Therefore, [48] consider a portfolio consisting of 32 loans of 120 currency units, 45 loans of 47 currency units and a remaining single loan of 45 currency units. We will refer to this portfolio as "concentrated portfolio".[9] The average sectoral concentration is supposed to be constant. Table 11.4 summarizes the results for EC_{sim} and its proxies EC^* and EC_{MFA} for both degrees of granularity.

Table 11.4. Comparison of EC^*, EC_{MFA} and EC_{sim} for portfolios with representative and low granularity (See [48], Table 10.)

Portfolio	Granularity	EC^*	EC_{MFA}	EC_{sim}	Rel. error of EC_{MFA}
Benchmark portfolio	representative	8.1	8.2	8.8	-7%
	low	8.0	8.0	9.3	-14%
Single sector portfolio	low	11.6	11.6	12.7	-8%

The results indicate that the EC_{sim} for the low granular benchmark portfolio is substantially higher than for the benchmark portfolio with high granularity considered in the previous subsection. In line with the results above, the EC_{sim} for the single sector portfolio is higher than for the benchmark portfolio. However, more important than the levels of economic capital are the errors of the economic capital approximations EC^* and EC_{MFA}. In this context, one should keep in mind that while the simulated economic capital takes granularity into account, EC^* and EC_{MFA} are calculated under the assumption of infinite granularity. The results in Table 11.4 show that the two proxies may underestimate the economic capital by up to 14%.

Finally, [48] investigate the impact of heterogeneous default probabilities within a sector on EC. Their results show that EC_{sim} decreases when using heterogeneous PDs due to the concavity of the dependence of EC on the default probability, while EC^* and EC_{MFA} remain unchanged since they do not account for heterogeneous PDs on obligor level. The EC approximations perform quite well for the representative benchmark portfolio. The approximation error for the single sector portfolio with low granularity amounts to +8% while it was −8% in the homogeneous case of Table 11.4. This indicates that PD heterogeneity dominates granularity effects. In summary, the results indicate that the effects of granularity and heterogeneity on the exposure level tend to counterbalance each other.

---

[8] See Directive 93/6/EEC of 15 March 1993 on the capital adequacy of investment firms and credit institutions.
[9] We already used this portfolio for comparison reasons in Section 9.4.4.

11.1.5 Discussion

The rules for determining minimum capital requirements for credit risk in the Basel II framework do not differentiate between portfolios with varying degrees of diversification. The Basel II framework does recognize the importance of potential concentration risk and requires this risk to be addressed in Pillar 2. In light of these observations, [48] in their empirical analysis show that economic capital increases with higher concentration of loans by business sectors and, therefore, underline the necessity to take concentration risk into account.

Given the finding that concentration of loans can substantially increase economic capital, it would be desirable to find analytical approximations for economic capital that avoid time-consuming Monte Carlo simulations. An analytical method for this purpose has been introduced by [109] and has been discussed in Section 10.1 above. The model allows to incorporate the effects of sector concentrations in a multi-factor framework. [48] evaluate the accuracy of this model for some realistic portfolios. The comparison of economic capital obtained from simulations and calculated within the multi-factor adjustment model shows that the analytical approximation performs very well for portfolios with homogeneous sectors and average granularity.

For banks and supervisors this result facilitates the determination of economic capital. Due to the good approximation quality of the measures described above, they can accurately approximate economic capital without conducting time-consuming Monte Carlo simulations. Moreover, the model of [48] takes care of some of the drawbacks of the approach by [28] as it allows for different asset correlations across sectors and it distinguishes between portfolios which are highly concentrated towards sectors with different degrees of correlation with other sectors.

Besides the positive implications of the findings above, some drawbacks of the analysis have to be addressed as well. First, some of the model assumptions are questionable. For instance, the analysis does not appropriately account for the impact of retail credits on the overall risk of a bank. Here we refer to the problem that, because for some banks, in particular small banks, corporate credits only account for a small share of the total portfolio, the capital effect for the bank should be weakened when incorporating retail credits. This is because the retail business exhibits only a small correlation with industry sectors. However, including retail exposures is rather difficult, since there is no representative stock index for them. This issue is a typical drawback of models which rely on index data for the estimation of factor correlations.

11.2 The Influence of Systematic and Idiosyncratic Risk on Large Portfolio Losses

In this section we review the work of [25]. The authors provide an empirical analysis of the influence of systematic and idiosyncratic risk on the portfolio loss distribution. Their analysis is based on data of banks' largest credit exposures from US regulators' Syndicated National Credits (SNC) examination programm. To account for dependencies between obligor's defaults, [25] use the multi-factor default-mode Merton model to estimate the inter- and intra-sector correlation parameters for each industry sector. The calibration procedure is based on KMV data for the Expected Default Frequencies (EDF) and asset returns of the borrowers.[10] They simulate the loss distribution of 30 real-world credit portfolios by means of Monte Carlo simulation. Based on these simulations, [25] examine the impact of expected loss, systematic risk and idiosyncratic risk on portfolio Value-at-Risk (VaR). Their empirical results show that, for very large Syndicated National Credit (SNC) portfolios, idiosyncratic risk has only a minor influence on VaR, while it can be quite significant for smaller portfolios. This result is in line with the empirical findings of [75] based on the granularity adjustment technique. On the contrary, the authors show that the contribution of systematic risk to portfolio VaR is almost independent of portfolio size.

[25], furthermore, investigate the extent to which credit portfolio VaR can be explained by simple indices of name and sector concentration. They show that simple concentration indices are positively correlated with portfolio VaR even after controlling for differences in average credit quality and portfolio size.[11] Moreover, they investigate the marginal contribution of individual sectors to portfolio VaR and its components.

11.2.1 Descriptive Analysis of SNC Data

[25] base their empirical study on data from US regulators' Syndicated National Credits (SNC) examination program. The SNC examination program was developed in 1977 and is supported by a database with annual updates. An exposure has to be reported to the SNC database when it exceeds 20 million US dollars and when it is held by three or more regulated entities. In their study, [25] analyze the SNC portfolios for the 30 largest lenders in 2005 measured by total dollar value of SNC commitments. These banks hold with US$760 billion about 46 percent of the total value of commitments in the database. Their exposures are grouped into 50 broadly-defined industry sectors[12] and the analysis is based on this sector-level data. As the data does

---
[10] Compare Section 3.3 for details on the KMV model.
[11] See [24], p. 1.
[12] Classification either by the North American Classification System or Standard Industrial Classification codes.

11.2 The Influence of Systematic and Idiosyncratic Risk   141

not include information on exposure distribution on obligor level, [25] assume the bank's commitments to a sector are all of the same size. This means that the exposure to a borrower in a particular sector k is the total commitment to that sector, divided by the number of loans in sector k. As in the previous section we denote by N the number of borrowers in the total portfolio and by N_k the number of borrowers in sector k. Then the exposure weight of sector k is given by $s_k = N_k/N$. To ensure privacy restriction on the bank-level data, [25] classify the 30 SNC portfolios into three different categories. These categories are *small portfolios* for portfolios with less than US$10 billion total exposure, *medium portfolios* for portfolios with between US$10 billion and US$20 billion of total exposure and *large portfolios* for portfolios with total exposure greater than US$20 billion. Moreover, they consider as a benchmark portfolio, for comparison reasons, the aggregated portfolio containing all syndicated national credits. In the following we will refer to this portfolio as the *All SNC* portfolio. We denote by s_k^* and N_k^* its sector exposure share and its number of obligors, respectively, for each sector $k = 1, \ldots, K$, where K denotes the number of sectors.

11.2.2 Simple Indices of Name and Sector Concentration

[25] study the degree to which simple ad-hoc measures[13] capture the impact of name and sector concentration on portfolio Value-at-Risk. To measure the amount of name concentration risk in the examined credit portfolios, they use the name concentration Herfindahl-Hirschman index

$$\text{HHI}_N = \sum_{k=1}^{K} \frac{s_k^2}{N_k} \in [0,1], \tag{11.2}$$

where K is the total number of sectors considered and the index N stands for name concentration. The index relates the weight on sector k measured by the sector's share of total commitments to the number of borrowers in sector k. This is also the reason why the index is denoted name concentration HHI. It is an HHI for the individual obligors in the portfolio when assuming that exposure shares are equally high within a sector. The index quantifies the deviation from an infinitely fine-grained portfolio with $\text{HHI}_N = 0$ and achieves a maximal value of $\text{HHI}_N = 1$ when the portfolio consists of a single exposure only. Alternatively, [25] suggest to measure name concentration relative to a well diversified real-world portfolio. Therefore, they introduce the *name concentration entropy index* e_N. The All SNC portfolio serves as a benchmark.

Definition 11.2.1 (Name Concentration Entropy Index)
The *name concentration entropy index* is defined as

---
[13] For a treatment of simple concentration indices and their properties see Chapter 8.

$$e_N = -\frac{\sum_{k=1}^{K} s_k \cdot (\ln(s_k/N_k) - \ln(s_k^*/N_k^*))}{\ln(\min_k(s_k^*/N_k^*))}. \tag{11.3}$$

$e_N = 0$ is achieved when all sector weights of the portfolio are the same as those in the All SNC portfolio. It approaches its maximum $e_N = 1$ in the limiting case where the portfolio weight of the obligor with the smallest weight in the All SNC portfolio dominates all others.

To measure sector concentration, another index relative to a real-world benchmark portfolio is used, the so called *sector concentration entropy index*. In contrast to the name concentration entropy index it does not depend on the number of obligors in each individual sector but only on the sector exposure shares.

Definition 11.2.2 (Sector Concentration Entropy Index)
The *sector concentration entropy index* is defined as

$$e_S = -\frac{\sum_{k=1}^{K} s_k \cdot (\ln(s_k) - \ln(s_k^*))}{\ln(\min_k(s_k^*))} \in [0,1]. \tag{11.4}$$

Here again $e_S = 0$ if the sector weights s_k coincide with those of the All SNC portfolio. It takes its maximum value one when all obligors belong to that sector with the smallest weight in the All SNC portfolio. For the sector concentration we analogously define a Herfindahl-Hirschman index as

$$\text{HHI}_S = \sum_{k=1}^{K} s_k^2 \in [0,1], \tag{11.5}$$

where the subscript S stands for sector concentration. Unfortunately this index is very sensitive to the sector definition, as, for example, aggregation of large sectors can dramatically increase the HHI_S.

11.2.3 Modeling Dependencies in Losses

Unfortunately the introduced indices cannot account for dependencies among obligors and their influence on name and sector concentration. Therefore, [25] simulate portfolio Value-at-Risk based on the multi-factor default-mode Merton model, introduced in Section 3.2. They also simplify this model using the multi-sector framework of Subsection 10.2.1 of which we adopt the notations here. The main problem in this framework is the estimation of the inter- and intra-sector correlation parameters.

From equation (10.23) in Subsection 10.2.1 we know that, for determining the correlation between the asset returns in case of the same sector, only an estimate of the intra-sector correlation β_k is needed whereas, in case of different sectors, also the cross-sector correlation parameters ϱ_k need to be estimated.

To estimate the intra-sector correlations β_k, [25] use KMV data on firms with average historical asset values exceeding US$500 million between 1993 and 2004. As explained in Section 3.3.1, KMV calculates the values of the firms' assets through a Merton based approach. [25] use these data to compute annual rates of asset returns r_n for every obligor n. Since the correlation of asset returns for obligors in the same sector depends only on the intra-sector correlation β_k, this is estimated as the square-root of the average empirical correlation in sector k.

In order to estimate the cross-sector correlations ϱ_k, normalized annual average rates of return for each sector k are used as proxies for Y_k and the empirical correlation among all proxy sector factors estimates the correlation matrix of the sector factors.[14] This leads to a standard problem of factor analysis and [25] solve it through an iterative principle components method.

The estimated inter- and intra-sector correlation parameters can be found in Table 5 in [24]. It can be noticed that the values for the intra-sector correlations vary strongly between the sectors. Moreover, in the principle components analysis for the derivation of the cross-sector correlations, [24] considered only 6 common risk factors with a very strong dependence on a single factor which might be interpreted as the overall economic condition. The other five factors reveal strong dependencies for similar sectors.

11.2.4 Monte Carlo Simulation of the Portfolio Loss Distribution

In order to simulate the loss distribution of the portfolios the following assumptions are made.

Assumption 11.2.3
(i) *All credit losses occur at the end of a one year horizon.*
(ii) *All credit losses occur only from obligor defaults.*
(iii) *The loss given defaults are non-stochastic and constant* ELGD $= 45\%$ *for all obligors.*
(iv) EAD $= 100\%$ *of the current exposure commitment amount for all obligors.*
(v) *The exposure shares of all loans in a given sector are equal.*
(vi) *The PDs of all obligors in the same sector equal the average KMV EDFs in that sector.*

Under these assumptions the loss rate L_k for sector k can be expressed as

$$L_k = \text{ELGD} \cdot \frac{1}{N_k} \cdot \sum_{n \in \text{Sector } k} D_n, \tag{11.6}$$

---
[14] Compare [24], p. 10.

where D_n is the default indicator of obligor n. Hence, $\sum_{n \in Sector\ k} D_n$ describes the number of defaults in sector k. As all obligors in one sector have the same default probability, the expected loss rate for sector k is

$$\mathbb{E}[L_k] = \mathrm{PD}_k \cdot \mathrm{ELGD}. \tag{11.7}$$

Denote the default threshold of any obligor in sector k by $c_k = \Phi^{-1}(\mathrm{PD}_k)$. From Chapter 3 we know that the default probability of obligor n in sector k given its sector factor Y_k equals

$$\mathrm{PD}_n(Y_k) = \Phi\left(\frac{\Phi^{-1}(\mathrm{PD}_k) - \beta_k Y_k}{\sqrt{1 - \beta_k^2}}\right). \tag{11.8}$$

Note that the conditional PD does only depend on the sector to which the obligor belongs. Thus, we replace the index n by the sector index k. Hence, the conditional expectation of the loss rate L_k for sector k is given by

$$\mathbb{E}[L_k | Y_k] = \mathrm{PD}_k(Y_k) \cdot \mathrm{ELGD}. \tag{11.9}$$

[25] express the loss rate for a portfolio of sector exposures as

$$L = \sum_{k=1}^{K} s_k L_k \tag{11.10}$$

$$= \underbrace{\sum_{k=1}^{K} s_k \cdot \mathbb{E}[L_k]}_{A} + \underbrace{\sum_{k=1}^{K} s_k \cdot (\mathbb{E}[L_k | Y_k] - \mathbb{E}[L_k])}_{B} + \underbrace{\sum_{k=1}^{K} s_k \cdot (L_k - \mathbb{E}[L_k | Y_k])}_{C}.$$

The portfolio loss is decomposed in three different terms to account for expected loss, systematic risk and idiosyncratic risk. Obviously, term A refers to the expected loss of the portfolio. Term B is accounting for losses due to systematic risk within and across sectors while term C measures the amount of idiosyncratic risk. The latter vanishes in case N_k tends to infinity for $k = 1, \ldots, K$.[15]

[25] simulate the portfolio loss distribution by means of Monte Carlo (MC) simulation. In each MC trial, standard normally distributed realizations of $\bar{X}$ and η_k are drawn. Together with the estimated cross-sector correlations ϱ_k, equation (10.22) then provides realizations for the sector factors Y_k. The default probabilities for each sector conditional on these realizations can then be obtained from equation (11.8) using the estimated intra-sector correlations β_k and the unconditional PDs given by the average KMV EDFs. The loss rate of each sector can be obtained from equation (11.6). Here [25] draw a realization of the number of defaults, $\sum_{n \in Sector\ k} D_n$, in sector k from a binomial distributed variable with probability $\mathrm{PD}_k(Y_k)$. Finally, the portfolio loss L can

---
[15] See e.g. Gordy (2003).

be simulated using equations (11.7), (11.9) and (11.10). For each portfolio 400.000 MC trials were conducted in order to simulate the loss distribution. Moreover, equation (11.10) can be used to investigate the impact of expected loss, systematic and idiosyncratic risk on the portfolio loss.

11.2.5 Empirical Results

[25] use the All SNC portfolio as well as three real bank portfolios of different sizes but with similar expected losses to investigate the effects of concentration risk on the loss distribution. The values for the different simulated components of credit risk for these portfolios are reported in Table 11.5 both as percentage of portfolio exposure and as percentage of portfolio VaR. Here VaR is computed as the 99.9% quantile of the loss distribution.

Table 11.5. Decomposition of VaR for selected portfolios (See [24], Table 8)

Portfolio	Risk Component	Percent of portfolio exposure	Share of VaR
All SNC	Expected loss	0.90%	19.05%
	Systematic loss	3.81%	80.25%
	Idiosyncratic loss	0.03%	0.69%
	VaR	4.75%	100%
Large Bank	Expected loss	0.88%	18.42%
	Systematic loss	3.86%	80.38%
	Idiosyncratic loss	0.06%	1.20%
	VaR	4.80%	100%
Medium Bank	Expected loss	0.88%	17.24%
	Systematic loss	4.07%	80.10%
	Idiosyncratic loss	0.14%	2.67%
	VaR	5.08%	100%
Small Bank	Expected loss	0.88%	16.34%
	Systematic loss	4.09%	75.74%
	Idiosyncratic loss	0.43%	7.92%
	VaR	5.40%	100%

The amount of expected losses is decreasing with decreasing portfolio size from 19.05% to 16.34% of portfolio VaR. Systematic risk is the largest component of portfolio VaR. It increases slowly from 75.74% to 80.25% of portfolio VaR with increasing portfolio size. Idiosyncratic risk is almost negligible for the All SNC portfolio with 0.69% but gets rather important for the small bank portfolios with 7.92%. This observation is in line with the empirical results of Section 9.1.

These results suggest a close relationship between portfolio size and degree of concentration. [25] support this claim by investigating the relation to the concentration indices introduced in Subsection 11.2.2. They calculated the concentration indices of Subsection 11.2.2 for the three real bank portfolios and for the All SNC portfolio. The results are reported in Table 11.6. One can clearly see, that the smaller the portfolio is, the higher are its name and sector concentration risk.

Table 11.6. Concentration indices for All SNC, large, medium and small portfolios (Compare [24], Table 6.)

Concentration Indices	All SNC Portfolio	Large Portfolios	Medium Portfolios	Small Portfolios
HHI_N	0.000	0.000	0.001	0.004
e_N	0.000	0.077	0.323	0.448
HHI_S	0.033	0.038	0.044	0.037
e_S	0.000	0.011	0.038	0.049

To examine the connection between the contributions to portfolio VaR and certain portfolio characteristics, [25] perform a regression of portfolio characteristics on unexpected losses (UL), including both systematic and idiosyncratic risk, for all 30 SNC portfolios in their data set using the concentration indices above as explaining factors. Their results indicate that total exposure is only significant (99%) if the concentration indices are not considered and the expected loss has always a significant (99%) positive influence on UL. For the concentration indices there are, as expected, positive relationships in the name-concentration HHI (99%), the name-concentration entropy (99%) and the sector concentration entropy (95%) index. It is interesting that the sector-concentration HHI is insignificant, which means that there is no relationship between this index and the unexpected losses[16] and that the sector-concentration entropy index is only significant at a 95% confidence level.

Finally, [25] also compute the marginal contributions of each sector to portfolio VaR. Therefore, note that the default rate $\frac{1}{N_k} \sum_{n \in Sector\ k} D_n$ on sector k in equation (11.6) can be expressed as

$$\frac{1}{N_k} \cdot \sum_{n \in Sector\ k} 1_{\{\beta_k Y_k + \sqrt{1-\beta_k^2}\varepsilon_n < c_k\}},$$

depending on the sector factor Y_k and the idiosyncratic shocks ε_n. Thus, the portfolio loss can be written as

$$L = \sum_{k=1}^{K} s_k \cdot ELGD \cdot \frac{1}{N_k} \cdot \sum_{n \in Sector\ k} 1_{\{\beta_k Y_k + \sqrt{1-\beta_k^2}\varepsilon_n < c_k\}}.$$

---

[16] This is a problem for the multi-factor adjustment model of [28] which uses this index.

11.2 The Influence of Systematic and Idiosyncratic Risk 147

Due to Lemma 1 in [76][17], the marginal contribution of sector k to total portfolio VaR can then be expressed as

$$\frac{\partial}{\partial s_k} \alpha_q(L) = \mathbb{E}\left[\text{ELGD} \cdot \frac{1}{N_k} \sum_{n \in \text{Sector } k} \mathbf{1}_{\{\beta_k Y_k + \sqrt{1-\beta_k^2}\varepsilon_n < c_k\}} \middle| L = \alpha_q(L)\right].$$

Based on this result, [25] apply Monte Carlo simulation to compute the marginal contributions of each sector to the total portfolio VaR. Due to equation (11.10), the marginal contributions can be split in expected loss, systematic risk and idiosyncratic risk. In their empirical analysis, the risk contributions for the All SNC portfolio and the three real-world portfolios are calculated based on six industrial sectors. Their results show that exposures to different sectors contribute very differently to VaR. An increase of an exposure in one sector might add very little to portfolio VaR while an equal increase in another sector can dramatically increase portfolio VaR. There are several reasons for this effect. First, the default probabilities between sectors vary. Hence a sector with a high PD obviously leads to higher VaR contributions than one with low PD. More importantly, however, correlation effects within sectors have a strong impact on the marginal VaR contributions. Increasing exposure in a sector with high intra-sector correlations leads to higher risk contributions than an equal increase of commitments to a sector with low intra-sector correlations. The empirical results of [25] demonstrate that a management of concentration risk depends on the structure of the loan portfolio. Especially, systematic risk has a much higher contribution to portfolio VaR than idiosyncratic risk. Hence, correlation effects have to be considered in any sensible risk management for loan portfolios. In sectors with low intra-sector correlation, reducing name concentration risk can be very effective while, for sectors with high intra-sector correlation, reducing sector concentration effects is more advisable.

11.2.6 Summary and Discussion

In summary, the main findings of [25] are

1. Simple name and sector concentration indices are positively correlated with portfolio VaR.
2. Systematic risk is the main part of portfolio VaR and nearly independent of portfolio size. Its marginal contribution to portfolio VaR depends not only on credit rating but also on the correlation structure of the sector loss rates.
3. Idiosyncratic risk in large portfolios is almost diversified away. But in small portfolios it plays a considerable role. Its marginal contribution to portfolio VaR is sector dependent and also dependents on the portfolio size.

---

[17] See also Lemma 9.1.1 in Section 9.1.

These results suggest that the most important task in managing a loan portfolio is the management of systematic risk. Especially in well diversified portfolios an increase in name concentration, in order to manage sector concentration due to correlation effects, will lead to a lower share of unexpected risk in portfolio VaR.

These findings underline the importance of risk concentrations in credit portfolio risk management and question the appropriateness of the ASRF model to calculate the risk weights for regulatory capital in Pillar I of the Basel II framework. The assumption of infinitely fine-grained portfolios, where idiosyncratic risk is diversified away, is only approximately fulfilled for large portfolios and cannot be held up for small credit portfolios. More critical is the share of systematic risk in the portfolio VaR and the marginal contribution of different sectors to portfolio VaR. [25] show that these influences result not only from the variation of average credit quality but also from the multi-factor dependence structure of credit losses showing that the assumption of a one-factor model is too simplistic due to correlation effects.

It might also be interesting to examine the influence of the obligor specific PDs and factor loadings on concentration risk. For example, the analysis of the marginal risk contributions suggests that high factor loadings in two of the sector factors could be a reason for a high marginal contribution to portfolio VaR. Moreover, empirical research about the nature of the six sector factors used in the analysis and a calculation of the portfolio sensitivity to them would give a deeper understanding about the dependence of the portfolio VaR on macroeconomic variables.

Part III

Default Contagion

12
Introduction

The examples mentioned in Chapter 7 as well as the empirical findings in Chapter 11 demonstrate that defaults are not stochastically independent across firms. Dependencies between different obligors in a credit portfolio can arise from two sources. First, we have *cyclical default dependence* resulting from dependencies of firms in a credit portfolio on some underlying macroeconomic factors as for example the overall economic climate. As these risk factors change in time, a firm's financial success, measured in terms of the default probability or its rating class, can vary. We refer to this source of risk as *systematic risk*. Methods to quantify the impact of systematic risk on a credit portfolio have been extensively discussed in Parts I and II. Most of these models rely on the conditional independence assumptions which says that, conditional on the underlying risk factors, the default events of the firms in a portfolio are independent. There exists a broad literature on the effect of cyclical default dependence in standard reduced-form models with conditional independence structure, see e.g. [46] or [89]. In these types of models, [81] and [115] showed that the conditional independence framework usually leads to default correlations between obligors which are too low to explain large portfolio loss that do occur from time to time. Hence as different channel of default dependence has to be considered.

Since common risk factors can be understood as a form of systematic risk effecting the whole economy, it is obvious to include, as another source of risk, counterparty relations. This means that we wish to incorporate in our models the effect that financial distress, initially effecting only a single obligor or a small group of obligors, can spread to a large part of the portfolio or even to the whole economy. An example of such an infectious chain reaction is the Asian crisis. Interactions which can be due to direct business links between firms, such as a borrower-lender relationship, can provide a channel for the spread of financial distress within an economic system or a portfolio. We refer to this source of default correlation risk as *default contagion* or *counterparty risk*. Default contagion leads to a higher default probability conditional on the default of another obligor than the unconditional default probability of a counterparty. The standard portfolio credit risk models we presented in Chapters 3-6 do not allow for default contagion through business links but rely on conditional independence among obligors. Thus, these models only take into account cyclical default dependence. As already mentioned in Definition 7.0.3, default contagion can increase the credit risk in a portfolio since the default of one borrower can cause the default of a dependent second borrower. Thus, default contagion can increase both name and sector concentration.

In this last part of these lecture notes we focus solely on concentration risk in credit portfolios that is due to counterparty risk. There has been an increasing interest in modeling default contagion in the recent literature. In the following chapters we will present and discuss some of the current models for default contagion starting in Chapter 13 with a discussion of work by [35], that provides empirical evidence for the importance of default contagion. The authors question whether common risk factors are the sole source underlying corporate default correlations and whether they can explain the extent of default clustering observed in empirical data. [35] find evidence of clustering beyond that, predicted by the conditional independence assumption, usually applied in reduced-form models. Their results question the ability of commonly applied credit risk models to fully capture the tail behavior of the probability distribution of portfolio losses. Thus, their research empirically underlines the importance of incorporating default contagion or clustering in an appropriate model for credit risk.

In the later chapters, we discuss different approaches to explicitly model default contagion in credit portfolios. Most of these models are formulated in the reduced-form setting. Note, that in reduced-form models default contagion has a direct impact on the default intensity of individual firms. If a business partner of firm n defaults, for example, the default intensity of the non-defaulted firm n jumps upwards. Reduced-form models can be divided in two different groups; copula models and models with interacting intensities. We present methods to incorporate contagion effects in both types of models.

In Chapter 14 we discuss the virtue of copulas for modeling default contagion, as they represent a useful tool to model dependencies of defaults of companies. Here the dependence structure between different obligors in a portfolio is specified a priori by the chosen copula. Hence, also the probability distribution of large portfolio losses is specified in advance. The individual default intensities are then derived from the copula structure. Although this approach might seem a bit unintuitive, copula models are very popular in practice since they can easily be calibrated to real data. [59] show that the correlation structure between companies does not contain enough information to capture the dependencies completely. To reduce the model risk, it is necessary to consider more relevant information by choosing an appropriate copula. Using Gauss copulas for instance, one faces the problem that they do not reflect certain

12 Introduction

properties noticeable in the data, as they are, for example, not able to capture fat tails. This is a drawback which affects a lot of existing credit risk models as, for example, the CreditMetrics and the KMV model. [59] state that the only adequate modification to capture this problem is to a find a copula being lower tail dependent for the latent variables, for example the t-copula or the Clayton copula. This idea is supported also by the fact that in large portfolios, quantiles of the tail of the mixing distribution are proportional to quantiles of the tail of the loss distribution.

In models based on interacting intensities the method is quite opposite to that in the copula approach. One first specifies, in which way the default of one firm might affect the default intensity of its business partners, and then derives the distribution of joint defaults or, respectively, the distribution of the portfolio loss variable. These models were first studied by [82] and [38]. The former developed a "primary-secondary" framework for pricing defaultable bonds in the presence of counterparty risk. In case a primary firm defaults, the spread of other obligors jumps upwards. Defaults of secondary firms do not have any effect on other firms in the portfolio. Unfortunately, this approach is computationally quite demanding. In the model of [38], the default of an obligor in the portfolio produces a regime switching from "low" to "high" risk in the sense that the default intensities of all obligors in the portfolio increase for a certain random time period. Afterwards the regime switches back to "low" again leading again to a jump in the default intensities. Other interacting intensity approaches that should be mentioned here have been introduced by [58] and by [49]. The former model presents an extension of the [82] approach, however, with the difference that default intensities depend on the average rating throughout the whole portfolio. This approach is also numerically quite complicated for larger portfolios. The model of [49] is better suited for practical application, however, with the drawback that it relies on numerical simulations as no analytical results can be derived.

Interacting default intensity models are closely related to models from interacting particle systems. One example of an application of interacting particle systems to credit risk is the voter-model based approach of [69] which we present in Chapter 15. The authors concentrate only on credit contagion by virtue of local interactions in a business partner network. These can, for example, be represented by borrowing and lending networks, as for example interbank lending or trade credits in the manufacturing sector. Due to the interdependence between borrowers, financial distress may spread through the whole economic system. In particular, the model at hand is a statistical one, not based on micro considerations. Hence it circumvents the complexity of typical micro models.

Chapter 15 also comprises an overview over some alternative approaches to model default contagion which are related to some extent to the model of [69] or which provide extensions of that model. We try to point out existing similarities and differences to demonstrate the relation more clearly and to emphasize the advantages of one or the other model.

There have also been attempts to model default contagion through incomplete information. [67] use a structural model setting where the default thresholds of the individual firms in the portfolio are not publicly available. Investors only obtain information on default thresholds whenever a firm defaults. Hence, the prior distribution, reflecting the publicly available information, has to be updated whenever new information arrives. Based on the updated posterior distributions, the authors model the evolution of the joint default probabilities in time. They observe that default probabilities exhibit a jump structure which reflects contagion effects.

Another incomplete-information approach to default contagion has been introduced by [63]. In contrast to the model of [67], the authors use a reduced-form model setting. They assume that the default intensities of the firms in a credit portfolio depend on some common state variable X, which is not directly observable. Investors can only observe historical default data of the portfolio and some noisy price observations of traded credit derivatives. Hence, they do not have full information on the distribution of the underlying state variable X. The approach of [63] is based on a two step procedure. First, they consider the model with full information on X. This can be solved with standard techniques for Markov processes. Then the general incomplete information problem can be understood as a projection of the full information model on the actually available information set. Hence, the main problem consists in the computation of the conditional distribution of X_t given the investors information at time t. This in turn can be understood as a nonlinear filtering problem.

In Chapter 15 we briefly discuss an approach of [33] which extends the ideas of these information-based contagion models, since it also considers the evolution of the indicators of financial distress in time.

Finally, in Chapter 16 we introduce an approach for financial contagion based on an equilibrium model. Here we review work by [80] who study an equilibrium model of credit ratings, where dependencies between different firms generate an intrinsic risk that cannot be diversified away and where small shocks, initially affecting only a small number of firms, spread by a contagious chain reaction to the rest of the economy. The arrival of an external shock impacts the economy through two channels; directly, via downgrades of individual credit ratings, and indirectly, through the deterioration of the average of credit ratings. The latter phenomenon is known as a "feedback effect", because the deterioration of the mean of credit ratings may produce a new wave of downgrades and, henceforth, can lead to even more financial distress economy wide. Moreover, [80] provide a method to quantify the additional risk arising from counterparty relations.

13
Empirical Studies on Default Contagion

One of the main challenges for the risk management of banks' portfolios remains to correctly assess the correlation of corporate defaults in order to allocate the right amount of economic capital to portfolio risk. Many banks apply credit risk models that rely on some form of the *conditional independence* assumption, under which default correlation is assumed to be captured by the dependence of all firms in the portfolio on some common underlying risk factors. Well known examples of this approach include the Asymptotic Single Risk Factor (ASRF) model as developed by the Basel Committee on Banking Supervision (BCBS) and [72], or applications of the structural Merton model [105] like the KMV or the CreditMetrics model.[1] Also reduced-form or mixture models, like the CreditRisk$^+$ model, rely on the conditional independence framework.[2] In the context of reduced-form models the conditional independence assumption is also often referred to as the *doubly stochastic* property. Essentially it says that, conditional on the paths of some common risk factors determining firms' default intensities, the default events of the individual firms are independent Poisson arrivals with (conditionally deterministic) intensities.

In the last years, the question arose whether correlations, generated by reduced-form models based on the doubly stochastic assumption, can sufficiently match observed real-world levels. One reason for a potential inability might be an insufficient specification of the underlying intensities. This problem has been discussed in the article [125].

In this chapter we review work by [35] who add some new evidence to this debate. In particular, the authors concentrate on the relatively little explored question of whether this class of models, with its known computational and statistical advantages, appropriately copes with the amount of *default clustering* observed in real world data. As already mentioned in Chapter 12, there are mainly two sources for default clustering. First, a firm's financial success

---
[1] See Chapters 3 and 4 for details on these models.
[2] See Chapters 5 and 6 for details on these models.

might be sensitive to common (macro-economic) factors. We refer to this effect as *cyclical default dependence*. Secondly, there might exist direct business links between firms in a portfolio which provide a channel for the spread of financial distress within a portfolio and thus leading to *default contagion* or *counterparty risk*. The first channel of default clustering can be modeled by the conditional independence framework. [35] test whether the dependence on macro-economic variables on its own is sufficient to explain correlated defaults in empirical data. The statistical tests in [35] are based on the fact that, under the doubly stochastic assumption, the process of cumulative defaults can be represented as a time-changed Poisson process. This provides some testable implications for the cumulative number of defaults in a portfolio. The authors suggest several possible tests and apply them to US data on corporate defaults from 1979 to 2004 using intensity estimates from [45]. Their results demonstrate that the data do not support the joint hypothesis of well specified default intensities and the doubly stochastic assumption, implying that the doubly stochastic model cannot explain the whole amount of default correlation in the data.

Hence, given the wide use of the conditional independence assumption through various kinds of credit risk models, the results of [35] indicate the need for a review of the currently applied portfolio credit risk models. This in turn may also have consequences for the allocation of regulatory capital and for rating and risk analysis of structured credit products which are exposed to correlated defaults in their underlying portfolios (e.g. collateralized debt obligations (CDOs) and options on portfolios of default swaps).

13.1 The Doubly Stochastic Property and its Testable Implications

In this section we first review some basic definitions and properties on counting processes where we follow mainly the work by [43]. Therefore, consider a probability space $(\Omega, \mathcal{F}, \mathbb{P})$ with filtration $(\mathcal{F}_t)_{t \geq 0}$ satisfying the usual conditions.

Definition 13.1.1 (Counting Process)
A stochastic process $(Z_t)_{t \geq 0}$ is called a *counting process* or *point process* if for a sequence of increasing random variables $\{\tau_0, \tau_1, \ldots\}$ with values in $[0, \infty]$, such that $\tau_0 = 0$ and $\tau_n < \tau_{n+1}$ whenever $\tau_n < \infty$, we have

$$Z_t = n \quad \text{for} \quad t \in [\tau_n, \tau_{n+1})$$

and $Z_t = \infty$ if $t \geq \tau_\infty = \lim_{n \to \infty} \tau_n$. The random variable τ_n represent the time of the nth jump of the process $(Z_t)_{t \geq 0}$. Hence, Z_t is the number of jumps that have occurred up to and including time t. The counting process is said to be *nonexplosive* if $T_\infty = \infty$ almost surely.

13.1 The Doubly Stochastic Property and its Testable Implications

Definition 13.1.2 (Stochastic Intensity)
Consider a non-negative and $(\mathcal{F}_t)$-predictable process $\lambda = (\lambda_t)_{t \geq 0}$ such that for all $t \geq 0$, we have $\int_0^t \lambda_s ds < \infty$ almost surely. Let $Z = (Z_t)_{t \geq 0}$ be a non-explosive adapted counting process. Then Z is said to admit the intensity λ if the compensator of Z admits the representation $\int_0^t \lambda_s ds$, i.e. if the process $(M_t)_{t \geq 0}$ defined by

$$M_t = Z_t - \int_0^t \lambda_s ds$$

is a local martingale.

Due to the martingale property of the process $(M_t)_{t \geq 0}$ associated with a non-explosive adapted counting process Z with intensity λ, we obtain

$$\mathbb{E}[Z_{u+t} - Z_t | \mathcal{F}_t] = \mathbb{E}\left[\int_t^{u+t} \lambda_s ds \bigg| \mathcal{F}_t\right].$$

Remark 13.1.3 A Poisson process is an example of an adapted non-explosive counting process Z with deterministic intensity λ such that, for all $t \geq 0$, we have that $\int_0^t \lambda_s ds$ is finite almost surely. Moreover, $Z_t - Z_s$ is Poisson distributed with parameter $\int_s^t \lambda_u du$ for all $s < t$.

Now consider a portfolio of N firms indexed by $n = 1, \ldots, N$. Suppose that the default time of firm n is described by the first jump time τ_n of a non-explosive counting process $(Z_t^n)_{t \geq 0}$ with stochastic intensity $(\lambda_t^n)_{t \geq 0}$.

Many methods to measure portfolio credit risk are based on a conditional independence structure. In the multi-factor Merton model, for example, one assumes that conditional on a set of systematic risk factors the individual loss variables of the obligors in the portfolio are independent. This framework is mathematically very convenient as we already saw in several of the discussed approaches for measuring credit risk and, in particular, for measuring concentration risk. However, the question arises whether the dependence on some common underlying factors can reflect the true dependence structure in a portfolio. In this chapter we want to answer empirically the question whether there exist interdependencies between obligors in a portfolio that go beyond the common dependence on some macro-variables. That is, we want to answer the question whether the joint distribution of the default times $(\tau_1, \ldots, \tau_N)$ in the consider portfolio is completely determined by the joint distribution of the intensities $(\lambda_1, \ldots, \lambda_N)$. If this is not the case, the conditional independence assumption underlying several credit portfolio models is questionable. That would imply that, even after conditioning on the paths of intensities of all firms, the default times can still be correlated. Therefore, let us start by reformulating the conditional independence assumption in the framework of stochastic intensities. This leads to the definition of doubly stochastic processes.

Definition 13.1.4 (Doubly Stochastic Processes)
Let $(Z_t)_{t\geq 0} = (Z_t^1, \ldots, Z_t^N)_{t\geq 0}$ be a multi-dimensional nonexplosive counting process with respective intensities $(\lambda_t)_{t\geq 0} = (\lambda_t^1, \ldots, \lambda_t^N)_{t\geq 0}$. Then Z is said to be *doubly stochastic*, if the following holds. Conditional on the path $(\lambda_t)_{t\geq 0}$ of all intensity processes and on the information $\mathcal{F}_T$ available at any given stopping time T, the counting processes $Z_{u+T}^1, \ldots, Z_{u+T}^N$ are independent Poisson processes with respective (conditionally deterministic) intensities $\lambda_{u+T}^1, \ldots, \lambda_{u+T}^N$.

We say that the stopping times $(\tau_1, \ldots, \tau_N)$ are *doubly-stochastic* with intensities $(\lambda^1, \ldots, \lambda^N)$ if the underlying multi-dimensional counting process, whose first jump times are $(\tau_1, \ldots, \tau_N)$, is double stochastic with intensities $(\lambda^1, \ldots, \lambda^N)$.

Now the conditional independence assumption, stating that default correlation is completely captured by the co-movement in intensities, can be expressed as follows.

Assumption 13.1.5 (Doubly Stochastic)
The multi-dimensional counting process $Z = (Z^1, \ldots, Z^N)$ is doubly stochastic.

Hence, given the intensities, the default times τ_n are independent, i.e. default correlation is completely driven by intensity correlation. Denote by

$$\bar{\lambda}_t \equiv \sum_{n \in \{1,\ldots,N\}:\ \tau_n > t} \lambda_t^n$$

the sum of the intensities of all non-defaulted firms at time t and let

$$K(t) = \#\{n : \tau_n \leq t\}$$

be the cumulative number of defaults by time t. Now consider a time-change $U(t)$ defined by

$$\frac{dU(t)}{dt} = \bar{\lambda}_t. \tag{13.1}$$

Then

$$U(t) = \int_0^t \sum_{n=1}^N \lambda_u^n \cdot \mathbf{1}_{\{\tau_n > u\}} du \tag{13.2}$$

is the cumulative aggregated intensity of surviving firms up to time t. The following proposition states a main consequence of the doubly stochastic assumption.[3]

Proposition 13.1.6 *Suppose Assumption 13.1.5 holds. Then the process*

$$J = \{J(s) = K(U^{-1}(s)) : s \geq 0\}$$

---
[3] See [35], page 99, for this proposition and its proof.

of the cumulative number of defaults by the new time $U^{-1}(s)$ is a Poisson process with rate parameter 1. Hence, for any $c > 0$, the successive numbers of defaults per time bin,

$$J(c), J(2c) - J(c), J(3c) - J(2c), \ldots,$$

are iid Poisson distributed with parameter c.

[35] use this result to test whether the data supports the doubly stochastic assumption. They divide their sample data into non-overlapping time bins containing an equal cumulative aggregate default intensity of c. The data would support the doubly stochastic Assumption 13.1.5 if the numbers of defaults in the successive bins are independent Poisson random variables with common parameter c.

13.2 Data for Default Intensity Estimates

The default intensity data used in [35] is taken from [45] who estimate the default intensity of firm n at time t from a non-linear regression by means of full maximum likelihood. The regression is based on two firm-specific co-variates, namely the distance-to-default of firm n and the trailing one-year stock return of firm n. Recall from Chapter 3 that the distance-to-default is an estimate for the number of standard deviations by which the assets of a firm exceed a measure of liabilities. Moreover, they use two macro-covariates, namely the 3-month U.S. Treasury bill rate and the trailing one-year return of the S&P 500 stock index.

Data on corporate defaults is obtained from Moody's Default Risk Service and CRSP/Compustat. Firm-specific financial data and the S&P 500 returns are from CRSP/Compustat databases. Treasury rates are from the web site of the U.S. Federal Board of Governors. [35] only consider those firms belonging to the "Industrial" category sector by Moody's and for which they also have a common firm identifier for the Moody's, CRSP and Compustat databases. This results in a total of 2,770 firms with monthly observations from January 1979 to October 2004, so overall 392,404 firm-months enter the data set. The data set includes 495 defaults. The number of defaults per month ranges from 0 to 12.

13.3 Goodness-of-Fit Tests

This section comprises a range of tests designed by [35] to examine whether the data set supports the implications of the doubly stochastic assumption given in Proposition 13.1.6. It should be emphasized here that a rejection of the doubly stochastic assumption can be due to two sources; its actual non-validity or due to misspecified intensities. Hence, the authors interpret each

13 Empirical Studies on Default Contagion

test as a test of the joint hypothesis of the doubly stochastic assumption and correctly specified intensities.

Given the estimated annualized default intensity λ_t^n for each firm n and each date t, the cumulative aggregated intensity $U(t)$ of all surviving firms up to time t is given by equation (13.2), where τ_n denotes the default time of firm n. [35] construct calendar times $t_0, t_1, t_2, \ldots$ with $t_0 = 0$ such that $U(t_k) - U(t_{k-1}) = c$ for all $k = 0, 1, 2, \ldots$. Hence the so constructed time bins contain exactly c units of accumulative default intensity each. The number of defaults in the k-th time bin is then denoted by

$$X_k = \sum_{n=1}^{N} 1_{\{t_k \leq \tau_n < t_{k+1}\}}.$$

Table 13.1. Comparison of empirical and theoretical moments for the distribution of defaults per bin (See [35], Table 1.)

Bin Size	Mean	Variance	Skewness	Kurtosis
2	2.04	2.04	0.70	3.49
(230)	2.12	2.92	1.30	6.20
4	4.04	4.04	0.50	3.25
(116)	4.20	5.83	0.44	2.79
6	6.04	6.04	0.41	3.17
(77)	6.25	10.37	0.62	3.16
8	8.04	8.04	0.35	3.12
(58)	8.33	14.93	0.41	2.59
10	10.03	10.03	0.32	3.10
(46)	10.39	20.07	0.02	2.24

The number K of bin observations is shown in parentheses under the bin size. The upper-row moments are those of the theoretical Poisson distribution under the doubly stochastic hypothesis; the lower-row moments are the empirical counterparts.

Table 13.1 compares the empirical moments of the distribution of the number of defaults X_k per time bin with the corresponding theoretical moments. If X_k is Poisson distributed with parameter c, i.e. $P(X_k = i) = e^{-c} \cdot c^i/i!$, then the associated mean and variance of X_k are equal to c, while the third moment amounts to $c^{-0.5}$ and the fourth moment equals $3 + c^{-1}$. The results in Table 13.1 show that the values for the empirical variance are bigger than those for the theoretical variance. Concerning skewness and kurtosis, the values for the empirical distribution differ from the corresponding theoretical moments more or less pronouncedly. [35] observe that the theoretical and empirical distribution are quite similar for smaller bin sizes, while they can differ substantially for bin sizes larger than 4.

[35] first test for correctly measured default intensities and the doubly-stochastic property using *Fisher's dispersion test*. This test compares the number of defaults X_k in the k-th time bin (for bin size c) with the theoretical Poisson distribution with intensity c. Under the null-hypothesis that $X_1, \ldots, X_K$ are independent Poisson distributed with parameter c, the random variable

$$W_K = \sum_{k=1}^{K} \frac{(X_k - c)^2}{c} \qquad (13.3)$$

is χ^2 distributed with $K-1$ degrees of freedom. Fix some $K \in \mathbb{N}$. Suppose the doubly stochastic Assumption 13.1.5 holds. Then, if the empirical values for W_K are large compared to the theoretical χ^2 distribution (with $K-1$ degrees of freedom), this would imply a large amount of defaults per time bin relative to the expected amount. [35] derive that, for all bin sizes, the hypothesis of correctly specified intensities and doubly stochastic defaults can be rejected at levels beyond 99.9%.

The second test, applied in [35], focuses on the tail behavior of the empirical distribution of the aggregated number of defaults per bin. If the upper tail of the empirical distribution is fatter than that of the theoretical Poisson distribution, this implies that the dependence between firms in the portfolio might not be fully captured by the conditional independence framework. The applied test, which is based on a Monte Carlo method, indicates that the upper-quartile tails of the empirical distribution are fatter than those of the theoretical Poisson distribution for 4 out of 5 bin sizes. This provides evidence for a source of default contagion that can not be explained by the doubly stochastic assumption.

Furthermore, [35] apply the so-called *Prahl test* which has been invented by [108]. Since this test can detect event clustering that cannot be explained by a Poisson distribution, it provides a further method to test whether the doubly stochastic assumption fully explains the dependence structure between firms in a credit portfolio. The test is based on the following reasoning.

Let $(\tau_1, \ldots, \tau_N)$ be, as before, the first jump times of the counting process $(Z^1, \ldots, Z^N)$. Now we apply the time change $U(t)$ defined in equation (13.1). Recall from Proposition 13.1.6 that the cumulative number of defaults by the new time U^{-1} are Poisson distributed with parameter 1. Denote the new jump times by

$$\theta_{n+1} := \inf \{s : J(s) > J(\theta_n)\}$$

with $S_0 = 0$, where the process J is defined as in Proposition 13.1.6. Due to Theorem 23.1 in [19], we know that the inter-jump times $\theta_{n+1} - \theta_n$ in this new time scale are iid exponential with parameter 1.

162   13 Empirical Studies on Default Contagion

Prahl's test is based on exactly this property. Table 13.2 shows the sample moments of inter-default times in both the actual time scale and the new time scale.[4] The results indicate that a large amount of default correlation in the portfolio can be explained by the doubly stochastic property, since the moments of the inter-arrival times in the new time scale (the intensity time scale) are much closer to the corresponding exponential moments than those of the actual inter-default times (in calender time). However, the test also implies that there can be excessive risk clustering such that the joint hypothesis of correctly specified conditional default probabilities and the doubly-stochastic assumption have to be rejected.

Table 13.2. Selected moments of the distribution of inter-default times (See [35], Table 4.)

Moment	Intensity time	Calendar time	Exponential
Mean	0.95	0.95	0.95
Variance	1.17	4.15	0.89
Skewness	2.25	8.59	2.00
Kurtosis	10.06	101.90	6.00

13.4 Discussion

As already mentioned before, there are mainly two sources for default clustering in time. Firms can be correlated due to dependence on some common underlying risk factors. Moreover, there might be direct business links between firms in a portfolio, also leading to an increase in default correlation. Most credit risk models which are applied in practice only take into account the first source of default contagion as they are only based on the conditional independence assumption. Thus, it is a straightforward question to ask whether these models are capable to reflect the true risk of default contagion in credit portfolios. The work of [35] focuses on exactly this problem. They test the empirical validity of whether common risk factors can completely explain the amount of default clustering observed in empirical data. Therefore, the authors focus, in particular, on the class of reduced-form models.

The main contribution is certainly the introduction of a novel time-changing technique. Assuming default times to be doubly stochastic, the process of cumulative defaults is Poisson distributed under a new time scale that is associated with the aggregate number of non-defaulted firms in the portfolio. This transformation allows to test the doubly stochastic assumption

---
[4] [35] have applied a linear scaling to equate the means of both time distributions.

13.4 Discussion

in a straightforward and simple way. The results in [35] indicate that the joint hypothesis of correctly measured intensities and the doubly-stochastic property can be rejected at standard significance levels. The authors also test whether this can at least partly be explained by the omission of significant covariates of measured intensities. Therefore, [35] include as macro-economic covariates both the growth rates of the U.S. gross domestic product (GDP) and the industrial production (IP). Thus the authors examine whether the additional inclusion of these covariates might help to explain default arrivals. For this purpose, [35] apply a linear regression model. Their results provide evidence that industrial production offers some explanatory power, but that GDP growth rates do not. Hence, the rejection of the doubly stochastic assumption in the above tests might be due to the omission of industrial production growth as a covariate. Therefore, [35] re-estimate default intensities to include IP growth. Using these new estimates for the default intensities, [35] repeat all their tests. The main conclusions, however, are the same as before, such that the estimated doubly stochastic model is rejected. Hence, it is questionable whether credit risk models, based solely on the conditional independence framework, can indeed reflect the actual tail behavior of portfolio losses.

Given the economic importance of risk capital allocation, it seems astonishing that the doubly stochastic assumption has not been tested in a similar context before. Hence, the work of [35] is highly valuable and may be of particular interest to risk managers and regulators. It might be necessary to re-think the amount of capital allocated to default correlation risk within standard industry credit risk models as well as within the Basel II framework which we presented in Chapter 4.

Some shortcomings of the analysis or implications for future research, however, might be addressed.

(i) Given that the authors consider the choice of correct intensities as crucial, different possible estimates should be discussed first and tests be conducted afterwards. This would provide a robustness check for the relative importance of intensities when testing the doubly stochastic assumption.
(ii) Taken at face value, results imply a re-consideration of existing models under the conditional independence assumption and make similar empirical tests of other credit risk models desirable.
(iii) In a former version of the paper, the authors use a shorter data span from 1987 to 2000 and KMV intensities. The doubly stochastic assumption is not as pronouncedly rejected as under the longer horizon. After using estimates of [45], the null hypothesis is not rejected any longer. Hence, further research might provide valuable evidence on whether the rejection is related to the specific time horizon and, if so, through which channels the rejection is related to the specific time horizon. Relating to this ob-

servation, it might make sense to subdivide the sample into shorter time intervals in order to check whether this changes results.

(iv) [35] only consider public U.S. firms. Given that a significant share of regulatory capital is also allocated to retail firms, the question of how results would change for non-public firms seems of interest. In particular, it seems plausible that the omitted variables problem, when estimating intensities, should be aggravated for smaller firms.

14

Models Based on Copulas

One of the main risks in the management of large credit portfolios is inherent in the occurrence of disproportionately many joint defaults of different counterparties over a fixed time horizon. Hence it is essential to understand the impact of the default correlations and the dependence structure between the different obligors in the portfolio on the portfolio loss variable. [61] study dependent defaults in large credit portfolios based on both latent variable models as well as mixture models. In the CreditMetrics and the KMV model (representing the latent variable models) the joint default probabilities are assumed to follow a multivariate normal distribution. By this assumption large joint losses are assigned to low probabilities. Empirical evidence, however, shows that the joint default probability of extreme losses is actually larger than that induced by a multivariate normal dependence structure. Examples are the oil industry where 22 companies defaulted in 1982-1986 or the retail sector where over 20 defaults occurred in 1990-1992[1]. Hence, assuming a normal distribution for portfolio losses, the probability of extreme losses can be underestimated. In order to reduce the model risk (the risk arising from a wrong specified model), [61] focus on the factors which have an impact on the tail of the loss distribution and, thus, also on the joint default probability as, for example, the default correlation and the individual default probabilities.

In latent variable models, default correlations are affected by asset correlations. However, as pointed out in [62], correlations are not sufficient to determine joint default probabilities. A useful tool to completely describe the dependence structure of defaults and, therefore, the joint default probabilities are copula functions. By copula functions the joint distribution can be separated into the marginal distribution functions themselves and the dependence structure of the joint distribution since copulas specify how the marginal distribution functions are combined to a joint distribution. A short introduction to copula functions is provided in the Appendix A. The KMV and CreditMet-

---

[1] See [113].

rics models, for example, are based on the Gaussian copula, as was observed by [90].

In this chapter we discuss the results of [61]. The authors explore the role of copulas in the latent variable framework and show that, for given default probabilities of individual obligors, the distribution of the number of defaults in the portfolio is completely determined by the copula of the latent variables. We present some simulation studies performed in [59] to give evidence for the theoretical results. The simulations indicate that, even for fixed asset correlations, assumptions concerning the latent variable copula can have a profound effect on the distribution of credit losses.

14.1 Equivalence of Latent Variable Models

Assume the general credit portfolio setting of Section 1.2. Recall the Definition 3.0.2 of a latent variable model $(V_n, (b_r^n)_{0 \leq r \leq R+1})_{1 \leq n \leq N}$ with marginals V_n and threshold values b_r^n from Chapter 3. Note that the distribution of the state indicator vector S is invariant with respect to simultaneous transformations of the marginals V_n and the threshold values b_r^n. Hence, there are many possibilities to set up latent variable models generating the same distribution of S. These models will be called equivalent.

Definition 14.1.1 (Equivalence of Models)
Consider two latent variable models with state vectors S and $\tilde{S}$. The corresponding latent variable models are called *equivalent* if the state vectors are equivalent in distribution.

The following theorem gives a sufficient condition to ensure the equivalence of two latent variable models.[2]

Theorem 14.1.2 *Let S and $\tilde{S}$ be the state vectors of two latent variable models $(V_n, (b_r^n)_{0 \leq r \leq R+1})_{1 \leq n \leq N}$ and $\left(\tilde{V}_n, (\tilde{b}_r^n)_{0 \leq r \leq R+1}\right)_{1 \leq n \leq N}$. The models are equivalent if the state vectors coincide in distribution ($S \stackrel{d}{=} \tilde{S}$), i.e. if the marginal distributions of S and $\tilde{S}$ coincide*

$$\mathbb{P}(V_n \leq b_r^n) = \mathbb{P}(\tilde{V}_n \leq \tilde{b}_r^n), \text{ for } r \in \{1, \ldots, R+1\} \text{ and } n \in \{1, \ldots, N\}$$

and if V and $\tilde{V}$ have the same copula.

The proof given in [59] uses Sklar's Theorem[3] and requirements for multivariate distributions. For $N = 2$ and distinguishing only between default and non-default, the result can be seen as follows.

---
[2] Compare [59], Proposition 3.3.
[3] See Theorem A.0.2 in the Appendix A.

14.1 Equivalence of Latent Variable Models

$$\mathbb{P}(S_1 = 0, S_2 = 0) = \mathbb{P}(V_1 \leq b_1^1, V_2 \leq b_1^2)$$
$$= C\left(\mathbb{P}(V_1 \leq b_1^1), \mathbb{P}(V_2 \leq b_1^2)\right)$$
$$= \mathbb{P}(\tilde{V}_1 \leq \tilde{b}_1^1, \tilde{V}_2 \leq \tilde{b}_1^2)$$
$$= \mathbb{P}(\tilde{S}_1 = 0, \tilde{S}_2 = 0)$$

where C is the common copula.

Using this result the models by KMV and CreditMetrics can be considered as equivalent as they both assume normally distributed individual default probabilities implying a Gaussian copula in both cases.

To study the impact of the copula of V on the loss distribution, [61] construct a subgroup of k obligors with individual default probabilities $\text{PD}_{n_1}, \ldots, \text{PD}_{n_k}$. Distinguishing only default or non-default states and applying Sklar's Theorem, we obtain

$$\mathbb{P}(D_{n_1} = 1, \ldots, D_{n_k} = 1) = \mathbb{P}\left(V_{n_1} \leq b_1^{n_1}, \ldots, V_{n_k} \leq b_1^{n_k}\right)$$
$$= C_{n_1, \ldots, n_k}(\text{PD}_{n_1}, \ldots, \text{PD}_{n_k})$$

where $C_{n_1, \ldots, n_k}$ denotes the corresponding k-dimensional margin of C. Assuming that V is exchangeable and thus that its copula is exchangeable, this equation reduces to

$$\pi_k = C_{n_1, \ldots, n_k}(\pi, \ldots, \pi), \tag{14.1}$$

where π_k denotes the joint default probability of an arbitrary subgroup of k obligors and π is the default probability of a single obligor which is identical for all counterparties in an exchangeable framework. Hence, the copula of V determines higher order joint default probabilities and thus the probability of many simultaneous defaults. An appropriate copula of V to model extreme risks is a copula that has lower tail dependence due to the fact that a low level of the assets leads to high default probabilities. Possible choices are the t-copula (see Example A.0.5), that provides both upper and lower tail dependence, and the Clayton copula which we will discuss in the example below. The Clayton copula belongs to the class of *Archimedean copulas*. These copulas can be specified by a so-called *generating function* or *generator*. They represent the distribution functions of exchangeable uniform random vectors and, thus, can only be used to characterize exchangeable latent variable models. The d-dimensional Archimedean copula is defined as

$$C(u_1, \ldots, u_d) = \theta^{-1}(\theta(u_1) + \ldots + \theta(u_d)) \tag{14.2}$$

where the generator function $\theta : [0, 1] \to [0, \infty]$ is a continuous, strictly decreasing function which satisfies $\theta(1) = 0$ and its inverse $\theta^{-1} : [0, \infty] \to [0, 1]$ is completely monotonic, i.e.

$$(-1)^m \cdot \frac{d^m}{dt^m}\theta^{-1}(t) \geq 0, \quad \text{for} \quad m \in \mathbb{N}.$$

Example 14.1.3 The generator function of the Clayton copula is given by $\theta_\kappa(u) = u^{-\kappa} - 1$ and the d-dimensional *Clayton copula* is, thus, defined by

$$C_\kappa^{Cl}(u_1, \ldots, u_d) = (u_1^{-\kappa} + \ldots + u_d^{-\kappa} + 1 - d)^{-1/\kappa}.$$

The Clayton copula exhibits lower tail dependence but upper tail independence. Therefore, it belongs to the class of *asymmetric copulas* meaning that these copulas possess asymmetric tail dependence. As an example of an Archimedean copula, the Clayton copula is only appropriate for exchangeable models. Consider, for example, a latent variable model where all individual default probabilities PD_n are equal to π. In this case we obtain the joint default probability π_k of an arbitrary subgroup of k obligors by applying equation (14.1), i.e.

$$\begin{aligned}\pi_k &= (\pi^{-\kappa} + \ldots + \pi^{-\kappa} + 1 - k)^{-1/\kappa} \\ &= (k\pi^{-\kappa} - k + 1)^{-1/\kappa}.\end{aligned}$$

Hence the joint default probability is completely determined by the constant individual default probability π and the parameter κ of the generator function of the Clayton copula.

14.2 Sensitivity of Losses on the Dependence Structure

We now concentrate on latent variable models as CreditMetrics and KMV with a factor model representation and with given fix asset correlation matrix R.[4] For fixed default probabilities, the Gaussian copula and the t-copula are chosen to examine the sensitivity of the credit loss distribution with respect to the selected copula. For the simulation study, the one-factor model representation is used where the obligor specific composite factors Y_n are given by

$$Y_n = \sqrt{\varrho} \cdot X$$

for a standard normally distributed variable X and where $\varrho \geq 0$. In the normal case, the asset value log-returns of obligor n can then be represented as

$$r_n = \sqrt{\varrho} \cdot X + \sqrt{1-\varrho} \cdot \varepsilon_n,$$

where ε_n is a standard normal idiosyncratic shock for obligor n which is independent of X.[5] This latent variable model can be represented by the exchangeable Gaussian copula C_ϱ^{GA} with parameter ϱ. In case of the t-distribution, we set

---

[4] See Section 5.3 for conditions when a latent variable model can be written as a Bernoulli mixture model.

[5] Compare Section 3.2 for this factor model representation.

14.2 Sensitivity of Losses on the Dependence Structure

$$r_n = \sqrt{\frac{\nu \varrho}{S}} \cdot X + \sqrt{\frac{\nu(1-\varrho)}{S}} \cdot \varepsilon_n$$

for a χ^2_ν-distributed random variable S independent of X and $\varepsilon_1, \ldots, \varepsilon_N$. Such a framework can be captured by an exchangeable t-copula $C^t_{\nu,\varrho}$ with ν degrees of freedom.

In the simulation studies, [59] vary the number N of obligors in the portfolio, the individual default probabilities, which are all identical to π in the exchangeable framework, the correlation of the latent variables ϱ and the degrees of freedom ν of the t-copula. Instead of simulating the latent variables directly, [59] use the corresponding Bernoulli mixture model representation with conditional default probabilities given by[6]

$$\pi = p_n(X) = \mathbb{P}(r_n < b_1^n | X) = \Phi\left(\frac{b_1^n - \sqrt{\varrho} \cdot X}{\sqrt{1-\varrho}}\right).$$

This corresponds to equation (3.8) with default threshold $b_1^n = \Phi^{-1}(\mathrm{PD}_n)$ which is constant equal to $\Phi^{-1}(\pi)$ in the exchangeable framework.

[59] separate obligors in three groups of decreasing credit quality, labeled by A, B and C. The default probabilities π and correlation parameters ϱ for the different groups are shown in Table 14.1 below.

Table 14.1. Values of π and ϱ for different obligor groups

	A	B	C
π	0.06%	0.50%	7.50%
ϱ	2.58%	3.80%	9.21%

Using 100,000 realizations of M, [59] compute the empirically estimated 95% and 99% quantiles of the distribution of the number of defaults M, denoted by $\alpha_{0.95}(M)$ and $\alpha_{0.99}(M)$, respectively. The results of the simulation study are summarized in Table 14.2. One can broadly recognize that, the lower the degree of freedom is, the higher the obtained defaults are. This is due to the upper tail dependence property of the t-distribution. [59] also point out that the quantile values are approximately proportional to the number N of obligors in the portfolio which reflects the result of Theorem 5.2.1.

The authors compare the quantiles for the Student t model with 10 degrees of freedom to those of the Gaussian model for Group B and a portfolio size of 10,000 (see Figure 1 in [59]). It can be recognized that, for high confidence levels, the quantiles for the Student t model are much higher than those of the Gaussian model; again reflecting the upper tail dependence property of the

---

[6] Compare Proposition 5.3.2 for the relation between both types of models.

t-copula. However, as the number ν of the degrees of freedom in the t-copula model has a dramatic influence on the quantile values, specifying only the correlation ϱ but not the parameter ν leads to high model risk.

Table 14.2. Estimated quantiles of the distribution of M (See [59], Table 1.)

N	Group	$\alpha_{0.95}(M)$				$\alpha_{0.99}(M)$			
		$\nu=\infty$	$\nu=50$	$\nu=10$	$\nu=4$	$\nu=\infty$	$\nu=50$	$\nu=10$	$\nu=4$
1000	A	2	3	3	0	3	6	13	12
1000	B	12	16	24	25	17	28	61	110
1000	C	163	173	209	261	222	241	306	396
10000	A	14	23	24	3	21	49	118	126
10000	B	109	153	239	250	157	261	589	1074
10000	C	1618	1723	2085	2587	2206	2400	3067	3916

14.3 Discussion

Copulas are a useful tool to model dependencies of defaults of companies. [59] show that the correlation structure between companies does not contain enough information to capture the dependencies completely. Taking only the correlation and the individual probabilities into account, the remaining model risk is still remarkable. To reduce this model risk, it is necessary to consider more relevant information on the dependencies between obligors by choosing an appropriate copula. Deciding, whether a copula is appropriate or not for a given latent variable model, can be achieved by means of calibration. Using Gaussian copulas, for instance, one faces the problem that they are not able to capture fat tails. This is a drawback which affects a lot of existing credit risk models as, for example, the CreditMetrics and the KMV model. An implementation of the changes in the specification of the credit risk models in order to address fat tails, would be interesting to see. [59] find out that the only adequate modification to capture this problem is to a find a copula representation of the latent variable model being lower tail dependent as, for example, the t-copula or the Clayton copula.

Although copula models are quite popular in practice, which is due to the fact that they are very easy to calibrate to prices of, for example, defaultable bonds or CDS spreads, they still have some shortcomings. In a copula model the dependence structure is exogenously specified from which the default intensities and the amount of default contagion are then endogenously derived. This is somewhat unintuitive. In the following chapters we study some approaches which are more intuitive in this aspect, however, they have of course some different shortcomings. The decision which drawbacks are more important, has to be decided from case to case. In practice, however, the decision is often driven by data availability and computational effort which again is an advantage for copula based models.

15
A Voter Model for Credit Contagion

In this chapter we want to emphasize the virtue of statistical mechanics approaches to socio-economics and, in particular, to the modeling of credit contagion phenomena. As it is shown in recent articles, for example by [22] in their famous binary choice models, there is a close relation between socio-economics and statistical mechanics. This motivates the use of interacting particle systems also in finance. The approach stems from the belief that the social or local structure among agents can be formalized. Local interactions are mediated by a social structure, not by the market. Moreover, this approach to local interactions is non-strategic but probabilistic. A general problem in these kinds of models – but not to be considered here – is that of *phase transition,* referring to the non-uniqueness of the ruling probability measure in the random economic system. In this sense, local data may not uniquely specify the global behavior of the system, as for example in the famous Ising model that is elaborately discussed in [66]. Moreover, these models necessitate a quite naive view of agents when paralleling models of the unlived nature in economics. This, however, can also constitute a great advantage as models from interacting particle systems do not require any rationality assumptions on agents.

Here, we review the model of [69] who encounter a *voter model* for credit contagion as a particular example of interacting particle systems in finance. The voter model is particularly adjuvant in modeling how an infectious process spreads among agents. It goes without saying that, in a broad literature in biology, voter models are used to explain the spread of diseases, as for example the chicken flu. In recent years, these types of processes have become more and more prominent in economics. It is for example in [87], where we see a model explaining the transmission of rumors through financial markets. We try to motivate the intuition behind the voter model as a quite natural device to explain contagious processes where states of agents are influenced by their peers' states. In particular, the model at hand is a statistical one, not based on micro considerations. Hence it circumvents the complexity of micro models as, for example, the one we discuss in Chapter 16.

In [69], the authors analyze the impact of default contagion on the portfolio loss distribution and, in particular, on the volatility of aggregate credit losses in large portfolios of financial positions. In this way, [69] investigate how default contagion increases the risk of large portfolio losses. The model under consideration constitutes a reduced-form contagion model where local interactions among firms are modeled statistically. The firms in a portfolio are identified with the nodes of a multidimensional lattice. Business partner relationships are characterized by the edges between different nodes. Thus each firm has a certain number of neighbors or business partners which depends on the dimension of the lattice. There are two possible liquidity states for each firm; *high* and *low* liquidity. Initially, a firm's state is random and the evolution of a firm's liquidity state in time follows a Poisson process. The intensity of the individual state transitions is proportional to the number of business partners in the low liquidity state. Hence, the more distressed business partners a firm has, the more likely the firm is to be distressed itself.

When the distribution of liquidity states is achieved by a voter model as in this situation, the loss of a firm is assumed to be random and dependent on its liquidity state. Given liquidity states, losses are assumed to be independent. In infinitely large portfolios, the average loss is governed by the mixing distribution that corresponds to the average proportion of low-liquidity firms, hence representing systematic risk. For practical applications, it is of course important to consider finite portfolios. Here, a Gaussian approximation to the loss distribution enables the measurement of aggregate portfolio risk. [69] show that contagion induces a higher probability of large losses. However, contagion effects decrease with increasing dimension and volatility of systematic risk.

15.1 The Model Framework

In this subsection we describe the voter model for credit contagion as introduced in [69]. The business partner network, determining the interactions between firms in a portfolio, is modeled by a neighboring structure on a multidimensional lattice. More specifically, firms are identified with their positions on the d-dimensional lattice $\mathbb{Z}^d$ which, thus, represents the index set of firms in the portfolio. It is assumed that firms interact with their business partners or neighbors. The set of neighbors of firm n is defined by

$$N(n) := \{m \in \mathbb{Z}^d : |n - m| = 1\} \subset \mathbb{Z}^d \setminus \{n\},$$

where $|\cdot|$ denotes the shortest path between two firms on the lattice $\mathbb{Z}^d$. Hence, each firm is identified with its location on the d-dimensional lattice $\mathbb{Z}^d$ and

interactions between firms are symmetric in the sense that
$$m \in N(n) \implies n \in N(m).$$
Moreover, all firms have the same finite number of business partners.

The crucial question now is, how do business partners influence each other? Therefore, each firm is assigned a liquidity state. Formally, the statistical model in [69] can be described by a probability space with underlying set
$$\Omega := \{0,1\}^{\mathbb{Z}^d},$$
the *(liquidity) configuration space*.[1] Here, $\{0,1\}$ denotes the *state space*, i.e. $\omega \in \Omega$ equips each firm $n \in \mathbb{Z}^d$ with a particular state and, hence, uniquely characterizes the state of the economy.

Definition 15.1.1 (Liquidity States)
For $\omega \in \Omega$, we say that firm n is in the *high liquidity state* if $\omega(n) = 0$ while $\omega(n) = 1$ corresponds to the *low liquidity state*.

When we consider the evolution of a firm's liquidity state in time, we will denote by $\omega_t(n)$ the state of firm n at time $t \geq 0$ such that $\omega_t \in \Omega$ for all times $t \geq 0$. We omit the subscript t when not considering the evolution of configurations in time.

It is assumed that the state of firm n depends statistically on the states of its business partners. Specifically, at a random time τ, a business partner $m \in N(n)$ has to pay a certain amount to its creditor m. If $\omega_\tau(m) = 0$, then m pays its debt, otherwise it defaults. Moreover, firm n is in the high liquidity state $\omega_t(n) = 0$ for times $t \geq \tau$ if and only if m fulfills its obligations, i.e. if $\omega_\tau(m) = 0$. Otherwise n is in the low liquidity state $\omega_t(n) = 1$ for $t \geq \tau$. Thus the liquidity state of the business partner m influences the state of the creditor n. If m pays its debt then n stays in the high liquidity state.

Assumption 15.1.2
The maturity time τ is an exponentially distributed random time with parameter one, i.e. $\tau \sim \exp(1)$.

For fixed $n \in \mathbb{Z}^d$ the business partner $m \in N(n)$, that has to pay its obligation, is chosen according to some distribution $p(n,m)$. For simplicity let $p(n,m) = 1/(2d)$ for all $n,m \in \mathbb{Z}^d$ with $|n-m| = 1$. Hence the business partner, whose payment is due at time τ, is chosen uniformly among all neighbors of firm n. Thus, the transition between liquidity states of firm n is given by a Poisson process with transition rate
$$c(n,\omega) := \sum_{m \in N(n)} p(n,m) \cdot |\omega(n) - \omega(m)|, \tag{15.1}$$

---
[1] For a more general introduction to configuration spaces and further notation, we refer to [66].

which in our case of uniform $p(n,m)$ equals

$$c(n,\omega) = \begin{cases} \dfrac{1}{2d} \displaystyle\sum_{m:|n-m|=1} \omega(m) & \text{if } \omega(n) = 0, \\ \dfrac{1}{2d} \displaystyle\sum_{m:|n-m|=1} (1-\omega(m)) & \text{if } \omega(n) = 1. \end{cases} \qquad (15.2)$$

The evolution of the firms' liquidity states in time can now be described by the process $(\omega_t)_{t \geq 0}$ with state space $\{0,1\}^{\mathbb{Z}^d}$ and transition rate c. Note that for arbitrary $n \in \mathbb{Z}^d$ the conditional probability distribution of the state $\omega_t(n)$ of firm n at time t given the whole history of firm n's liquidity states up to and including time $s \leq t$ (represented by the σ-algebra $\mathcal{F}_s$), depends only on the state $\omega_s(n)$ of the process at time s. Hence, the process $(\omega_t)_{t \geq 0}$ satisfies the Markov property and the associated Markov transition function $(P_{s,t})_{s,t \geq 0}$ satisfies

$$P_{s,t} P_{t,u} = P_{s,u}$$

for $s < t < u$, where

$$\mathbb{E}\left[f(\omega_t) | \mathcal{F}_s\right] = P_{s,t}(\omega_s, f) \quad \text{almost surely}$$

for any continuous function f on Ω. The process $(\omega_t)_{t \geq 0}$ is time homogeneous, such that we obtain a transition semigroup P_t given by

$$P_{t-s} = P_{s,t}$$

for $s \leq t$. It can be shown that, for any bounded and continuous functions f on Ω and for each $t > 0$, the function $P_t f = \mathbb{E}[f(\omega_t)|\mathcal{F}_0]$ is continuous on Ω. This is known as the *Feller property*. Hence, the evolution of the firms' liquidity states in time can be described by a continuous time Feller process $(\omega_t)_{t \geq 0} = (\omega_t(n))_{n \in \mathbb{Z}^d, t \geq 0}$ with state space $\Omega = \{0,1\}^{\mathbb{Z}^d}$ and transition rate c. If all states align, we have $c = 0$, meaning that all firms stay in their original liquidity state. We refer to the evolution mechanism, described by saying that firm n changes from state $\omega(n)$ to $|1 - \omega(n)|$ at rate $c(n,\omega)$ given by equation (15.2), as the *voter model dynamics*.

Intuitively, a high liquidity firm switches its state at a rate proportional to the number of its low-liquidity neighbors. Consider a high liquidity firm in a trade credits context. As the number of business partners in the low liquidity state increases, the probability of a payment default increases as well. Hence, the firm is more likely to change its state to low liquidity, too.

15.2 Invariant and Ergodic Measures for the Voter Model

Let μ be the initial distribution of the evolution process $(\omega_t)_{t\geq 0}$ of a firm's liquidity state and assume that μ is translation-invariant.[2] We denote the distribution of the process $(\omega_t)_{t\geq 0}$ with initial distribution μ by $\mathbb{P}^\mu$. Then, $\mathbb{P}^\mu_t$ denotes the marginal distribution of ω_t under $\mathbb{P}^\mu$. Let

$$\varrho := \mu\{\omega_0 \in \Omega : \omega_0(n) = 1\} \qquad (15.3)$$

denote the Bernoulli parameter of the *initial marginal liquidity distribution* for an arbitrary firm n, i.e. ϱ measures an individual firm's marginal liquidity risk. Note that, since μ is assumed to be translation-invariant, every firm has the same marginal risk, initially.

Let $\mathcal{P}(\Omega)$ denote the set of probability measures on Ω with the weak topology. Then we can define the following properties for measures in $\mathcal{P}(\Omega)$.

Definition 15.2.1 (Invariant Measures)
A measure $\mu \in \mathcal{P}(\Omega)$ is said to be *invariant* for the process with Markov semigroup $(P_t)_{t\geq 0}$ if $\mu P_t = \mu$ for all $t \geq 0$ where $\mu P_t \in \mathcal{P}(\Omega)$ is defined by

$$\int f d(\mu P_t) = \int P_t f d\mu$$

for all continuous functions f on Ω.

μP_t is interpreted as the distribution at time t of the process when the initial distribution is μ. Hence, for the process $(\omega_t)_{t\geq 0}$ the measure μP_t corresponds to the distribution $\mathbb{P}^\mu_t$. Let $\mathcal{I}$ denote the set of invariant measures on Ω for the voter model, i.e. under the Markov semigroup associated with the Feller process $(\omega_t)_{t\geq 0}$. Then $\mathcal{I}$ is convex and closed in the weak topology. We denote by $\mathcal{I}_{ex}$ the set of extremal elements in $\mathcal{I}$ which are defined as follows.

Definition 15.2.2 (Extremal Element)
A measure $\mu \in \mathcal{I}$ is called an *extremal element* of $\mathcal{I}$ if μ is not a proper convex combination of other elements of $\mathcal{I}$.

It can be easily seen, that the extremal elements in the set of translation-invariant measures correspond to the ergodic measures in $\mathcal{P}(\Omega)$, where the latter are defined as follows.

---

[2] *Translation invariance* generalizes the notion of *stationarity* for stochastic processes to the multidimensional case. For $\omega \in \Omega$ and $n \in \mathbb{Z}^d$ we define the translation $T_n(\omega)(m) = \omega(n+m)$. Then T_n canonically operates on the subsets of Ω. We call a measure μ on Ω translation-invariant if $\mu(A) = \mu(T_n A)$ for all $n \in \mathbb{Z}^d$ and for all measurable sets $A \subset \Omega$.

Definition 15.2.3 (Ergodicity)
(1) A translation-invariant distribution $\mu \in \mathcal{P}(\Omega)$ is called *ergodic*, if μ is trivial on the σ-algebra of translation-invariant sets; meaning that $\mu(A) \in \{0, 1\}$ for all translation-invariant subsets $A \subset \Omega$.
(2) The Markov process with semigroup $(P_t)_{t \geq 0}$ is said to be *ergodic* if the set of all invariant measures for the process is a singleton $\mathcal{I} = \{\nu\}$ and if

$$\lim_{t \to \infty} \mu P_t = \nu$$

for all $\mu \in \mathcal{P}(\Omega)$.

Note that the voter model has two trivial invariant measures, δ_0 and δ_1, where δ_i, for $i \in \{0, 1\}$, denotes the Dirac measure on the constant configuration with $w(n) = i$ for all $n \in \mathbb{Z}^d$. Thus the voter model is not ergodic in the sense of Definition 15.2.3 (2). It can be shown that, in case $d \leq 2$, there are no further extremal invariant measures for the voter model while, in the case $d \geq 3$, there is a one-parameter family

$$\mathcal{I}_{ex} = \{\nu_\varrho : 0 \leq \varrho \leq 1\}$$

of extremal invariant measures, where ν_ϱ is translation-invariant and ergodic, and $\nu_\varrho\{w : w(n) = 1\} = \varrho$. In the limit, as $t \to \infty$, the following theorem holds.[3]

Theorem 15.2.4 *Let $\Omega = \{0, 1\}^{\mathbb{Z}^d}$ and let μ be any translation-invariant initial distribution with Bernoulli parameter $\varrho = \mu\{w_0 : w_0(n) = 1\}$.*

1. *For $d \leq 2$ the marginal distribution of w_t under $\mathbb{P}^\mu$ converges weakly, as $t \to \infty$, to a convex combination of the two trivial invariant measures,*

$$\mathbb{P}_t^\mu \longrightarrow \varrho \delta_1 + (1 - \varrho)\delta_0.$$

2. *Suppose μ is ergodic. For $d \geq 3$ the distribution $\mathbb{P}_t^\mu$ of w_t converges weakly to the non-trivial extremal invariant measure ν_ϱ,*

$$\mathbb{P}_t^\mu \longrightarrow \nu_\varrho$$

with parameter $\varrho = \nu_\varrho\{w : w(n) = 1\}$.

Hence, as it is often the case when considering Markov processes, the dimension d of the index set $\mathbb{Z}^d$ plays an important role and we have to discuss the two cases, $d \leq 2$ and $d \geq 3$, separately. We begin in the next subsection with the *non-dense business partner network* where $d \leq 2$. For the *dense business partner network*, where $d \geq 3$, the analysis strongly relies on a result on the decomposition of a translation-invariant measure into a mixture

---
[3] For a more elaborate discussion of this result we refer to [92], Corollary 1.13 in Chapter V.

of ergodic measures. Therefore, we denote by $\mathcal{P}_e(\Omega)$ the space of all ergodic probability measures on Ω equipped with the weak topology. Denote by $\mathcal{G}$ the Borel σ-algebra on $\mathcal{P}_e(\Omega)$. Let $\mathcal{P}_{e,\varrho}(\Omega)$ be the subspace of $\mathcal{P}_e(\Omega)$ of probability measures ν with $\nu\{\omega : \omega(0) = 1\} = \varrho$. Note that $\nu\{\omega : \omega(0) = 1\} = \varrho$ also implies that $\nu\{\omega : \omega(n) = 1\} = \varrho$ for all $n \in \mathbb{Z}^d$, as $\nu \in \mathcal{P}_e(\Omega)$ is translation-invariant. The famous theorem of Choquet (compare Theorem 14.9 in [66]) establishes a decomposition of translation-invariant measures in ergodic ones, i.e. we can write any translation-invariant measure μ on Ω as the barycenter of ergodic measures with respect to some probability measure γ on $\mathcal{P}_e(\Omega)$,[4]

$$\mu = \int_{\mathcal{P}_e(\Omega)} \nu \gamma(d\nu).$$

[69] prove the following *refined ergodic decomposition*.[5]

Theorem 15.2.5 *Let $\mu \in \mathcal{P}(\Omega)$ be translation invariant. Then there exists a probability measure $\mathbb{Q}$ on $[0,1]$ and a kernel*

$$\gamma_\cdot(\cdot) : \begin{cases} \mathcal{G} \times [0,1] \to [0,1] \\ (A, \varrho) \mapsto \gamma_\varrho(A) \end{cases}$$

with $\gamma_\varrho(\mathcal{P}_{e,\varrho}) = 1$ such that

$$\mu = \int_0^1 \int_{\mathcal{P}_e(\Omega)} \nu \gamma_\varrho(d\nu) \mathbb{Q}(d\varrho). \tag{15.4}$$

15.3 The Non-Dense Business Partner Network

In this subsection, we are mainly concerned with the asymptotics of the liquidity process $(\omega_t)_{t \geq 0}$ and the properties of its *equilibrium distribution* in a non-dense business partner network where $d \leq 2$. Therefore, recall the following definition.

Definition 15.3.1 (Equilibrium Distribution)
$\mu \in \mathcal{P}(\Omega)$ is called an *equilibrium distribution* if μ is invariant under the voter model dynamics.

In the case $d \leq 2$, we then obtain from Theorem 15.2.4 of the previous subsection that

$$\mathbb{P}^\mu_t \longrightarrow \varrho \delta_1 + (1-\varrho)\delta_0,$$

---

[4] Note that γ is a distribution on the set of distributions $\mathcal{P}_e(\Omega)$, not a distribution on Ω.
[5] Compare Theorem A.2 in [69].

weakly as $t \to \infty$. Moreover, we observe a (global) clustering behavior, i.e. for all $n, m \in \mathbb{Z}^d$ we have

$$\lim_{t \to \infty} \mathbb{P}^\mu [\omega_t(n) \neq \omega_t(m)] = 0.$$

The intuition behind these results is immediate. In the long-run, only one type of firms survives. With probability ϱ all firms are in the low liquidity state. Marginal risk does not change under contagion, but it stays ϱ. Hence, we see invariance on the microscopic level but possibly drastic changes on the macroscopic level. For the trade credit case with ϱ relatively large, high liquidity firms become infected pretty easily and stable clusters of low liquidity firms merge. Eventually, with probability ϱ, all firms are in the low liquidity state.

15.4 The Dense Business Partner Network

Now consider a *dense network* where $d \geq 3$. In this situation we want to make use of the decomposition of translation-invariant initial distributions into a mixture of ergodic ones given by Theorem 15.2.5. Therefore, we first study the equilibrium behavior in the ergodic case. Thus, assume μ being ergodic. In this case we obtain from Theorem 15.2.4 that the marginal distribution $\mathbb{P}^\mu_t$ of ω_t under $\mathbb{P}^\mu$ converges weakly to the extremal invariant measure ν_ϱ with parameter $\varrho = \nu_\varrho \{\omega : \omega(n) = 1\}$ as $t \to \infty$, i.e. we have

$$\mathbb{P}^\mu_t \longrightarrow \nu_\varrho$$

weakly for $t \to \infty$. Intuitively, we now have coexistence of clusters in the long run, i.e. the contagion process is non-trivial. We obtain a fixed proportion ϱ of low-liquidity firms in the portfolio in the long run.

Let us now consider a general (not necessarily ergodic) translation-invariant initial distribution μ. In this case, the distribution $\mathbb{P}^\mu_t$ of $(\omega_t)_{t \geq 0}$ converges weakly to a mixture of the extremal invariant measures ν_ϱ ($\varrho \in [0,1]$) of the voter model. The key is to consider the distribution of the empirical proportion of low-liquidity firms $\bar{\varrho}$ as defined below.[6] To see that this term is well-defined, we refer to [79].

Definition 15.4.1 (Empirical Proportion of Low-Liquidity Firms)
Let μ on Ω be translation-invariant. The *empirical proportion of low-liquidity firms* $\bar{\varrho}$ is a random variable that is μ-almost surely defined as

$$\bar{\varrho} := \lim_{N \to \infty} \frac{1}{|\Lambda_N|} \sum_{n \in \Lambda_N} \omega(n),$$

where $\Lambda_N := \{n \in \mathbb{Z}^d : -N \leq n_l \leq N, l = 1, \ldots, d\}$.

---
[6] Compare [69], Definition 3.2.

15.4 The Dense Business Partner Network

Applying Theorem 15.2.5 to our situation, we can now formulate the main result in case of a dense business partner network (Theorem 3.3. in [69]).

Theorem 15.4.2 *Let $d \geq 3$ and let μ be translation-invariant on Ω. Suppose μ is given in its refined ergodic decomposition (15.4). Then the following holds*

$$\mathbb{P}_t^\mu = \int_0^1 \int_{\mathcal{P}_e(\Omega)} \mathbb{P}_t^\nu \gamma_\varrho(d\nu) \mathbb{Q}(d\varrho),$$

and we have

$$\mathbb{P}_t^\mu \longrightarrow \int_0^1 \nu_\varrho \mathbb{Q}(d\varrho) =: \mathbb{P}_\infty^\mu$$

weakly as $t \to \infty$, where ν_ϱ denotes the extremal invariant measure of the basic voter model in dimension $d \geq 3$ with parameter $\varrho \in [0,1]$ as introduced above.

The refined ergodic decomposition may be seen as a two stage stochastic process. First, we choose ϱ according to $\mathbb{Q}$, which in turn is given by Theorem 15.2.5. Then this prescribes the corresponding translation-invariant distribution

$$\mu_\varrho = \mu_{\varrho,0} := \int_{\mathcal{P}_e(\Omega)} \nu \gamma_\varrho(d\nu).$$

The same intuition holds true for the liquidity distribution $\mathbb{P}_t^\mu$ and $\mathbb{P}_\infty^\mu$ in that their decomposition works completely analogously. In particular, the latter is the barycenter of the extremal invariant measures ν_ϱ of the voter model with respect to the distribution $\mathbb{Q}$ of the (initial) empirical distribution $\bar{\varrho}$. The next result is due to [69], Corollary 3.4.

Corollary 15.4.3 *Given the assumptions of Theorem 15.4.2, the following holds for the empirical proportion $\bar{\varrho}$ of low-liquidity firms.*
(1) *$\bar{\varrho}$ is $\int_{\mathcal{P}_e(\Omega)} \mathbb{P}_t^\nu \gamma_\varrho(d\nu)$-a.s. equal to ϱ for $\varrho \in [0,1]$ and $t \in [0,\infty)$.*
(2) *For $t \in [0,\infty]$ the law of $\bar{\varrho}$ under $\mathbb{P}_t^\mu$ equals $\mathbb{Q}$.*

The first part of the corollary strongly reminds us of ergodic theorems as for example in [79]. We see that the distribution of low-liquidity firms is invariant under the contagion dynamics, i.e. it is not affected by the interaction of firms but it is fixed at ϱ. However, as it is discussed in [66], local interaction causes dependence across liquidity states. As locally contagious interactions are introduced, for any finite number of firms, the probability to observe a majority of firms with identical liquidity states has increased compared to the case of independent firms. This is what [66] calls the "breakdown of the law of large numbers". However, a priori, it is not known which state the majority takes over.

15.5 Aggregate Losses on Large Portfolios

In this section we concentrate on the aggregate losses in a portfolio and the impact of the credit contagion phenomenon on large portfolio losses in the case $d \geq 3$. We only consider the equilibrium case, that is the case where the distribution of the firms' liquidity states is invariant under the voter model dynamics. We consider a financial institution holding a portfolio of positions issued by firms $n \in \Lambda_N \subset \mathbb{Z}^d$ for some $N \in \mathbb{N}$, where Λ_N is defined as in Definition 15.4.1 Note that the number of firms in such a portfolio equals $(2N+1)^d$. Each position is subject to credit risk. Depending on its liquidity state, an obligor may not be able to pay an obligation. We are particularly interested in the distribution of the bank's aggregate losses.

Let the random variable L_n denote the loss on positions contracted with firm n. Then the portfolio loss is given by

$$L_{|\Lambda_N|} = \sum_{n \in \Lambda_N} L_n.$$

The index $|\Lambda_N|$ refers to the number of obligors in the portfolio, which here equals $(2N+1)^d$. We impose the following assumptions on the loss variables.

Assumption 15.5.1
Individual losses L_n are independent conditional on the liquidity state $r \in \{0,1\}$ of a firm, i.e. the conditional distribution $\mathbb{M}_r$ of losses with respect to a firm in state r only depends on the state r. Furthermore, losses are supported in a bounded interval on $\mathbb{R}_+$.

Given the distributions $\mathbb{M}_r$, we denote the expected loss, caused by a firm in state r, by

$$\mathrm{EL}_r = \int x \mathbb{M}_r(dx).$$

Intuitively, it is straightforward to suppose that expected losses for firms in the low liquidity state are higher than for those in the high liquidity state, i.e. we suppose $\mathrm{EL}_1 > \mathrm{EL}_0$. The following simplifying assumption on the conditional distribution of losses will be relaxed later on.

Assumption 15.5.2
Assume $\mathbb{M}_r = \delta_r$ for $r \in \{0,1\}$. In particular, this implies that L_n depends deterministically on the state r; there is no loss in the high liquidity state but unit loss in the low liquidity state.

Let $\mu = \int_0^1 \nu_\varrho \mathbb{Q}(d\varrho)$ be an equilibrium distribution, where ν_ϱ are the extremal invariant measures of the voter model for $0 \leq \varrho \leq 1$ and $\mathbb{Q}$ is the distribution of the random empirical proportion $\bar\varrho$ of low liquidity firms. Applying an ergodic theorem on the extremal invariant distribution ν_ϱ of the voter model, we obtain the following result for the average loss $L_{|\Lambda_N|}/|\Lambda_N|$ in the portfolio[7]

$$\lim_{N\to\infty} \frac{L_{|\Lambda_N|}}{|\Lambda_N|} = \bar{\varrho} \quad \mu\text{-almost surely.}$$

Here we used that, due to Assumption 15.5.2, the conditional loss distribution $\mathbb{M}_r$ of a firm in state r only depends on r. Thus, given a firm n in state r, the conditional expected loss of firm n equals

$$\mathbb{E}[L_n|\omega(n) = r] = r.$$

Hence, the distribution of average losses in large portfolios is solely determined by the distribution of the proportion $\bar{\varrho}$ of low-liquidity firms.

Remark 15.5.3 Since $\bar{\varrho}$ is a non-trivial random variable, we see that not all uncertainty averages out. It remains uncertainty of losses given by the distribution $\mathbb{Q}$, representing the systematic risk in the portfolio. Note that the average portfolio loss is not governed by the interactions between firms in the portfolio but by the distribution $\mathbb{Q}$ of the empirical proportion $\bar{\varrho}$ of low liquidity firms in the portfolio. To sum up, contagion does not increase the average loss, but it increases the risk of large losses in finite portfolios.

Let us first constrain ourselves to the case $\mathbb{Q} = \delta_\varrho$ for fixed $\varrho \in (0,1)$. Then the last remark can be formalized by the following theorem.[8]

Theorem 15.5.4 *Let $d \geq 3$ and $\mathbb{Q} = \delta_\varrho$ for $\varrho \in (0,1)$. Suppose Assumption 15.5.2 holds. For large portfolios the distribution of total losses $L_{|\Lambda_N|}$ can be approximated by a normal distribution $\mathcal{N}$, i.e.*

$$|\Lambda_N|^{-\frac{d+2}{2d}} \cdot (L_{|\Lambda_N|} - |\Lambda_N| \cdot \varrho) = |\Lambda_N|^{-\frac{d+2}{2d}} \cdot \sum_{n \in \Lambda_N} (\omega(n) - \varrho) \longrightarrow \mathcal{N}(0, \sigma^2)$$

weakly as $N \to \infty$, where the limiting variance $\sigma^2 = \sigma^2(d)$ is given by

$$\sigma^2 = \varrho \cdot (1-\varrho) \cdot \frac{\gamma_d \cdot d}{2^{d+3} \pi^{\frac{d}{2}}} \cdot \Gamma\left(\frac{d-2}{2}\right) \int_{[-1,1]^d} \int_{[-1,1]^d} \frac{1}{||x-y||_2^{d-2}} dx dy.$$

Here Γ denotes the Gamma-function and $\gamma = \gamma_d$ is given by

$$\frac{1}{\gamma} = \frac{1}{(2\pi)^d} \int_{(-\pi,\pi)^d} \left(1 - \frac{1}{d} \sum_{m=1}^d \cos x_m\right)^{-1} dx.$$

The loss distribution can be uniformly approximated by

---

[7] More precisely, since we assume conditional independence, we may apply a conditional law of large numbers.
[8] Compare [69], Theorem 4.1.

$$\sup_{x \in \mathbb{R}_+} \left| \nu_\varrho(L_{|\Lambda_N|} \geq x) - \Phi\left(\frac{|\Lambda_N|^{\frac{1}{2}} \cdot \varrho - |\Lambda_N|^{-\frac{1}{2}} \cdot x}{\sigma \cdot |\Lambda_N|^{\frac{1}{d}}} \right) \right| \leq \varepsilon_N,$$

where $\varepsilon_N \to 0$ as $N \to \infty$, and Φ denotes the standard normal distribution function.

The last part of Theorem 15.5.4 allows us to uniformly approximate the probability that the loss of a portfolio of size u exceeds $x \in \mathbb{R}_+$, and hence to approximate quantiles of the portfolio loss distribution.

Corollary 15.5.5 *Under the same conditions as in Theorem 15.5.4, the probability of a loss larger than $x \in \mathbb{R}_+$ for a given portfolio of size u can be uniformly approximated by*

$$\Psi_{d,\varrho}(u, x) = \Phi\left(\frac{u^{\frac{1}{2}} \cdot \varrho - u^{-\frac{1}{2}} \cdot x}{\sigma(d) \cdot u^{\frac{1}{d}}} \right),$$

i.e. $L_u \sim \mathcal{N}(\varrho \cdot u, \sigma^2(d) \cdot u^{1+\frac{2}{d}})$ *approximately.*

In case of independent firms, it can be shown that $L_u \sim \mathcal{N}(\varrho u, \varrho(1-\varrho)u)$ approximately. Hence the order of the variance term, which measures the risk of large portfolio losses, is u in case of independent firms and it depends on the dimension d in the contagion case where it equals $u^{1+\frac{2}{d}}$. Thus, contagion increases the risk of a portfolio (order u vs. order $u^{1+\frac{2}{d}}$) and riskiness decreases as d increases.

Remark 15.5.6 It should be emphasized, that contagion does not change the average loss but only the risk of large portfolio losses. The average loss stays the same (ϱu) whatever d is or whether firms are independent or not. However, the loss distribution exhibits a higher variance in case of contagion, i.e. portfolios are more risky in terms of large losses when firms interact locally. The probability of losses exceeding a given level above the average loss is increased, the lower d is, since volatility decreases in d. This can be interpreted by the fact that a denser network (i.e. high d) induces less "face-to-face"-interactions in the sense that the degree of interactions is lower, the higher d is. Moreover, the lower the Bernoulli parameter ϱ of the initial marginal liquidity distribution, the lower is the likelihood of large losses.

We now relax the assumption $\mathbb{Q} = \delta_\varrho$, but assume $\mathbb{Q}(\{0\}) = \mathbb{Q}(\{1\}) = 0$ to exclude situations where all firms are in the same liquidity state. The probability of a loss larger than $x \in \mathbb{R}_+$ is given by

$$\int \nu_\varrho(L_{|\Lambda_N|} \geq x) \mathbb{Q}(d\varrho).$$

Approximation of this probability is achieved by an appropriate mixture of normal distributions.[9]

---
[9] Compare [69], Corollary 4.2.

Proposition 15.5.7 *Let $d \geq 3$ and let $\mathbb{M}_r = \delta_r$ for $r \in \{0,1\}$. The distribution of portfolio losses $L_{|\Lambda_N|}$ can uniformly be approximated as*

$$\sup_{x \in \mathbb{R}_+} \left| \int \nu_\varrho(L_{|\Lambda_N|} \geq x) \mathbb{Q}(d\varrho) - \int \Phi\left(\frac{|\Lambda_N|^{\frac{1}{2}} \cdot \varrho - |\Lambda_N|^{-\frac{1}{2}} \cdot x}{\sigma(\varrho) \cdot |\Lambda_N|^{\frac{1}{d}}} \right) \mathbb{Q}(d\varrho) \right| \leq \varepsilon_N,$$

where $\varepsilon_N \to 0$ as $N \to \infty$.

Similarly to Corollary 15.5.5 one can derive an approximating function for quantiles of the portfolio loss distribution in this situation.

[69] then extend their theory to the general case, i.e. without Assumption 15.5.1 on the shape of the conditional distribution $\mathbb{M}_r$ to be in force. Hence conditional losses are stochastic in this situation. As before, let the initial distribution $\mu = \int_0^1 \nu_\varrho \mathbb{Q}(d\varrho)$ be an equilibrium distribution. First, we consider the average loss in the portfolio Λ_N. The joint distribution of losses is given by the mixture

$$\beta(dx) = \int \left(\bigotimes_{n \in \mathbb{Z}^d} \mathbb{M}_{\omega(n)} \right)(dx) \mu(d\omega), \quad x \in \mathbb{R}^{\mathbb{Z}^d}.$$

Again by the conditional law of large numbers, the following is β-a.s. well-defined

$$\lim_{n \to \infty} \frac{L_{|\Lambda_N|}}{|\Lambda_N|} = \varrho \cdot (\mathrm{EL}_1 - \mathrm{EL}_0) + \mathrm{EL}_0,$$

where EL_r is the expected loss on a position with a firm in liquidity state $r \in \{0,1\}$. The next proposition states that in large portfolios, the quantiles $\alpha_q(L_{|\Lambda_N|})$ of the portfolio loss distribution are governed by those of $\mathbb{Q}$.[10]

Proposition 15.5.8 *Denote by $\alpha_q(\mathbb{Q})$ the qth quantile of the distribution $\mathbb{Q}$ and assume that the cumulative distribution function of $\mathbb{Q}$ is strictly increasing at $\alpha_q(\mathbb{Q})$. Then*

$$\lim_{N \to \infty} \frac{\alpha_q(L_{|\Lambda_N|})}{|\Lambda_N|} = \alpha_q(\mathbb{Q}) \cdot (\mathrm{EL}_1 - \mathrm{EL}_0) + \mathrm{EL}_0,$$

where $\alpha_q(L_{|\Lambda_N|})$ denotes the qth quantile of the distribution of $L_{|\Lambda_N|}$ under the measure β.

Intuitively, the proposition says that in infinitely large portfolios the tail behavior of the portfolio loss L_∞ is essentially determined by the tail properties of $\mathbb{Q}$ and hence by the systematic risk in the portfolio. This result coincides with Proposition 4.5 of [59] which we recalled in Chapter 5, Theorem 5.2.1. In that case, the loss of firm n was given as

---

[10] Compare [69], Proposition 4.3.

$$L_n = \mathrm{LGD}_n \cdot D_n,$$

where D_n denotes the default indicator variable of firm n. Hence, there is only loss in the default case, such that EL_0 in the above Proposition would be zero. In the situation of Theorem 5.2.1, the portfolio was assumed to be homogeneous in the sense that all firms have the same ELGD. Thus in the situation of Chapter 5, the expected loss of firm n, conditional on a systematic risk factor X, can be described by

$$\mathrm{ELGD} \cdot p(X),$$

where $p(X)$ denotes the conditional default probability. Hence, EL_1 in the above Proposition would correspond to the expectation ELGD of the random variable of loss given default and $\alpha_q(\mathbb{Q})$ would correspond to $\alpha_q(p(X))$.

Remark 15.5.9 The law β of losses $L_{|A_N|}$ can again be approximated by a normal distribution. As before, [69] first constrain to the case $\mathbb{Q} = \delta_\varrho$ with $\varrho \in (0,1)$. In this case the expected loss is given by $\varrho(\mathrm{EL}_1 - \mathrm{EL}_0) + \mathrm{EL}_0$. From Theorem 15.5.4, the authors derive the weak convergence of the losses in the stochastic case. The result in the case of stochastic conditional losses is basically the same as before. Only the limiting variance is multiplied by a factor $(\mathrm{EL}_1 - \mathrm{EL}_0)^2$. Moreover, the averages ϱ of low liquidity states are replaced by $\varrho(\mathrm{EL}_1 - \mathrm{EL}_0) + \mathrm{EL}_0$.

15.6 Discussion and Comparison with Alternative Approaches

[69] present a default contagion model based on the voter model. Their approach is a quite simplified representation of the contagion mechanism in credit portfolios. Several assumptions should be rethought and eventually relaxed. For example, firms are assumed to be homogeneous. In particular, all firms possess the same marginal risk of default. This results from the assumption that the initial distribution μ is translation-invariant. A possible way out might be a Gibbsian approach with different reference measures across firms. Also different firm sizes could be considered such that the impact of a default would vary across firms.

A second shortcoming of the model is the symmetry of local interactions. The assumption, that the direction in which shocks are propagated is symmetric, might not be realistic for practical applications. Usually a larger company is less affected by financial distress of the smaller business partner than vice versa. This could be circumvented by considering directed graphs as in several percolation models, e.g. in [4] or [57]. The approach of [49], which we present later on in Section 15.7, also incorporates directed graphs, such that ideas of that approach might provide useful extensions of the model discussed in this

15.6 Discussion and Comparison with Alternative Approaches

chapter. Alternatively, instead of making use of the voter model, asymmetric interactions could be introduced by means of contact processes as introduced in [92].

Moreover, the model assumes that maturity times of different borrowers are distinct. It is not clear what would happen to the liquidity state of an obligor if two of his business partners default at the same maturity time.

In [68], the authors obtain a unifying theory by incorporating cyclical correlations via macro-variables in the present model. This approach provides a method to quantify the aggregate portfolio loss resulting not only from contagion through business partner networks but also from systematic risk through common dependence on underlying macro-economic factors. In this sense, the approach in [68] is comparable to the model of [49] as both take care of the two channels of default correlation; default contagion and cyclical default dependence. As already mentioned above, the model of [49] is more general in the sense that it can reflect a richer interaction structure through the use of directed graphs. For comparison reasons we discuss the latter model in some detail in Section 15.7 below.

Let us now briefly present some alternative models for default contagion which are related to the one discussed in this chapter but which we will not study in detail here. We try to point out existing similarities and differences and discuss some advantages and drawbacks of one or the other model.

15.6.1 The Mean-Field Model with Interacting Default Intensities

[58] present a model with interacting intensities where the default indicator process of the firms in a portfolio is modeled as a finite-state Markov chain conditional on some underlying macro-economic variables. The states represent the default state of the obligors at a given point in time and the transition rates characterize the default intensities. [58] apply an explicit probabilistic model such that the default indicator process can only jump to the neighboring states. The authors derive an explicit form for the conditional transition rates in terms of forward-equations which can be easily solved numerically for small portfolio sizes N. For large N the equations are, however, no longer useful and one needs to either reduce the dimension of the state space or one has to resort to simulation approaches. [58] suggest a dimension reduction method by considering a homogeneous group structure in the sense of a *mean-field* model. Therefore, it is assumed that one can divide the portfolio of N firms into K homogeneous groups in the sense that the risks within a group are exchangeable. Each group can, for example, be identified with firms in the same rating category. Moreover, it is assumed that default intensities of firms belonging to the same group are identical. Thus, the losses, which a firm might cause, only depend on the firm's rating category. This homogeneity implies that default intensities are invariant under permutations, which leave

the group structure invariant, and that default intensities of different firms from the same group are identical. This so-called *mean-field* model is a natural counterparty-risk model for portfolios consisting of homogeneous groups. For general portfolios, however, this approach might seem rather restrictive. Note, however, that it is the aim of the approach in [58] to study the impact of default contagion on large portfolio losses. Thus, it can be argued that in the limit, as the portfolio size tends to infinity, the homogeneity assumption is not very restrictive anymore. A prominent example which also makes use of this property, is the ASRF model underlying the Basel II framework. Furthermore, the restriction can be easily relaxed by increasing the number of homogeneous groups, however, with the drawback of also increasing the computational effort. Hence, the problem with this approach is to find a good balance between a realistic group structure and reasonable computational effort.

By applying Markov process techniques, [58] then investigate the behavior of the aggregate portfolio loss in the model with homogeneous groups as the portfolio size increases. They assume thereby that the number K of groups remains unchanged. The authors prove that in the limit the evolution of the proportion of defaulted firms in each group becomes deterministic given the evolution of the macro-economic factor. Thus the proportion of defaulted firms is fully determined by the evolution of the economic factors as N tends to ∞. This coincides with a result in [61] which we discussed in Chapter 5 and which is also similar to the result of [69] presented in Proposition 15.5.8.

Finally, [58] present some simulation results which demonstrate the impact of counterparty risk on default correlations and quantiles of the proportional defaults in the affine mean-field model with counterparty risk. Their results show that the default correlations and quantiles of the loss distribution increase substantially when the amount of interaction in the portfolio is increased. Note that this also corresponds to a result in [69], reformulated in Corollary 15.5.5. The latter states that the probability of losses exceeding a given level is increased when decreasing the parameter d. Recall that low values for d refer to a higher proportion of "face-to-face" interactions and thus to a higher amount of counterparty risk.

As already mentioned a shortcoming of the general Markovian model is that it becomes numerically intensive as the number N of firms in the portfolio increases. For large N, usual simulation methods are even superior to this approach. The suggested method of homogeneous groups circumvents this problem and provides a natural setting for modeling counterparty risk when firms in a portfolio are classified by means of rating categories for example. The number of homogeneous groups, however, still has to be quite small for this method to be efficient. The theoretical result coincides with results obtained in other approaches we presented in this part of these lecture notes. For these

15.6 Discussion and Comparison with Alternative Approaches 189

reasons and, because of the mathematically quite demanding methods used in that approach, we have discussed this model only very briefly here.

15.6.2 A Dynamic Contagion Model

Another model for the propagation of financial distress in a credit portfolio with locally interacting firms has been introduced recently by [33]. As the other approaches discussed in this chapter, it is also based on methods from interacting particle systems. The authors present an interacting intensities framework in a reduced-form model setting where the default probabilities of the individual firms in the portfolio depend on the states of the other obligors. The model accounts for two channels of default correlations. On the one hand, contagion effects are modeled through direct interactions due to business partner relationships. On the other hand, cyclical default dependencies are described by indirect interactions due to information effects for firms in the same sector. The idea behind such an information driven default model is the following. Macro-economic variables influencing all firms in a portfolio are not necessarily perfectly observable for all firms. Hence, the default probabilities of the obligors in the portfolio are influenced differently depending on there information. This effect is sometimes also called *frailty* and leads to an increase in default correlations. In the context of the conditional independence assumption, usually applied to model cyclical default dependencies, this means that the distributions of the underlying factors are not known to all firms in a portfolio. Say, for example, a factor influencing firms in the energy sector is only observable for those firms in that sector while firms belonging to other sectors do not have access to information on this factor. This idea has also been discussed in work by [114], [44] or [47]. Hence, the model of [33] generalizes attempts of, for example [49] which we present in the next Section 15.7, and which also incorporates both channels of default correlation, however, under the assumption of perfect information on the underlying macro-variables.

[33] study the impact of default contagion on large portfolio losses by analyzing the limiting loss distribution of homogeneous portfolios. Similarly to the model of [58], their model also makes use of the simplifying assumption of a homogeneous group structure in the sense of mean-field interactions. This assumption has already been discussed in some detail in the previous subsection. A major innovation of the approach in [33], compared to the already discussed ones, is that the authors present a dynamic model in the sense that it is possible to describe the evolution of the indicators of financial distress in time. Hence the impact of default contagion on large portfolio losses can be studied for any time horizon T. This also allows to study the asymptotic behavior of portfolio losses when the number N of firms in the portfolio tends to infinity and $T \to \infty$.

A shortcoming of the approach of [69], already mentioned in the introduction to this chapter, is that the phenomenon of phase transition is not considered. Phase transition refers to the effect that the aggregate state of a system is not uniquely determined by local data.[11] Here, phase transition is eminent in case of clustering; we know a priori that spins align but we do not know, which phase eventually emerges. The dynamic model setting of [33] allows to investigate the effect of phase transition. Therefore, denote the loss, a bank may suffer at time T from a portfolio consisting of N firms, by $L_N(T)$, which is computed as the sum of the individual losses of all firms in the portfolio at time T. Theorem 4.1 in [33] states that the difference between the actual portfolio loss $L_N(T)$ and the asymptotic loss $L_\infty(T)$, for $N \to \infty$, can be approximated by a normally distributed random variable with mean 0 and variance depending on the interaction structure and on time T. This result is similar to Theorem 15.5.4, however, in a much more general framework as the result in [33] allows to study the time behavior of large portfolio losses. Thus, phase transition can be studied in this framework.

15.7 Contagion Through Macro- and Microstructural Channels

As already mentioned several times in the previous Subsection 15.6, the model introduced by [49] provides an extension of the model by [69] in several ways. Most importantly, it accounts not only for default contagion but it also takes care of cyclical default dependencies. Moreover, it provides a much richer interaction structure than the one in the model of [69]. Therefore, we will devote this section to a short review of their approach where we will also point out the main differences to the other methods discussed in this chapter.

[49] interpret correlations between the firms in a portfolio, which are due to the dependence on some common underlying factors, as *macro-structural* dependencies. The authors model the macro-structure by a typical latent variable model similar to the multi-factor Merton model we discussed in Section 3.2. Further interdependencies between debtors, as for example direct business links, are represented by the *micro-structure* which provides a possible source for contagion effects. Micro-structural effects are modeled by using weighted graphs where the nodes represent the firms and the edges characterize business partner relationships between firms. Each edge is associated with a weight that represents the strength of the business partner relationship. Incorporating micro-structural information in the latent variable framework leads to a change in the idiosyncratic terms. Based on this extended latent variable framework, it is possible to quantify the impact of default contagion on the portfolio loss distribution. The results of [49] show that counterparty

---

[11] For economic systems, this phenomenon is elaborately discussed in [54].

15.7 Contagion Through Macro- and Microstructural Channels

risk increases the probability of large portfolio losses which coincides with the results obtained in previously discussed models of this chapter.

15.7.1 A Model with Macro- and Micro-Structural Dependence

As the model of [49] incorporates micro-structural dependencies between the obligors of a portfolio in a typical multi-factor asset value model that takes care of the macro-structural dependencies, we start in this subsection with a summary of the underlying latent variable model. We consider a portfolio of N firms indexed by $n = 1, \ldots, N$. To account for sectoral effects, [49] group the debtors into a small number K of different sectors. The dependence on macro-structural variables is expressed by assuming a conditional independence structure.

Assumption 15.7.1
Asset returns of different debtors are independent conditional on some random vector X.

The underlying factor vector X is assumed to be given by systematic sector-specific risk factors

$$X = (X_k)_{k=1,\ldots,K} \sim \mathcal{N}(0, \Sigma), \quad \Sigma = (\sigma_{k,l})_{k,l=1,\ldots,K}.$$

Assume that each obligor is only influenced by the risk factor of the sector to which it belongs. Hence, we can express the synthetic asset return r_n for obligor n by a univariate standard Gaussian latent variable

$$r_n = \sqrt{1 - w_{k(n)}^2} \cdot X_{k(n)} + w_{k(n)} \cdot \varepsilon_n, \quad r_n \sim \mathcal{N}(0,1), \qquad (15.5)$$

where $k(n)$ denotes the sector to which obligor n belongs. Here ε_n represents an idiosyncratic factor specific to obligor n. The ε_n's are standard normally distributed for $n = 1, \ldots, N$ and mutually independent. The sensitivity of obligor n's asset return to this idiosyncratic factor is specified by the weight $w_{k(n)}$ depending on the sector $k(n)$ to which obligor n belongs.

Up to now, the model corresponds to a typical multi-factor asset value model as, for example, the multi-factor Merton model we discussed in Chapter 3. The innovation of [49] is to introduce additional micro-structural dependencies to account for business partner relationships. Therefore, we represent the strength of the business interdependence between two firms in the portfolio by a *business matrix*

$$\Xi = (\xi_{ij})_{i,j=1,\ldots,N}.$$

As firms cannot interact with themselves, we assume that $\xi_{ii} = 0$ for $i = 1, \ldots, N$. Moreover, define a vector

$$\eta = (\eta_i)_{i=1,\ldots,N}$$

which attaches weights to the residual risk of each debtor. [49] now define the micro-structure for a credit portfolio of N obligors as follows.[12]

Definition 15.7.2 (Micro-Structure)
A *micro-structure* for a portfolio of N obligors is a directed weighted graph $\mathcal{G} = (N, \mathcal{E}, \Xi, \eta)$ where the nodes correspond to the counterparties. A directed weighted edge $\mathcal{E}_{nm}$ indicates a business relation from m to n that induces a counterparty risk for firm n with strength ξ_{nm} given by the edge weights. The node weight η_n represents the residual risk of debtor n.

Take, for example, two firms n and m in different sectors. When applying the macro-structural framework of equation (15.5), then firm n is influenced by $X_{k(n)}$ but it cannot a priori be influenced directly by $X_{k(m)}$ and vice versa for firm m. The only interdependence between firms n and m is generated through the covariance matrix Σ of the sector factors. Due to the interdependencies on the micro-structural level represented by the business relation ξ_{nm}, however, firm n can be influenced by $X_{k(m)}$ and also by the idiosyncratic risk ε_m of firm m. Moreover, the residual firm-specific idiosyncratic risk for firm n reduces to η_n.

Remark 15.7.3 This setting generalizes the model of [69] in several ways. First, it allows for asymmetric effects since ξ_{nm} does not necessarily equal ξ_{mn}. Moreover, not every firm has the same number of business partners in this setting. Recall that in [69] the set of business partners of firm n was simply defined as the set of vectors $m \in \mathbb{Z}^d$ with distance one to the vector $n \in \mathbb{Z}^d$. In the model of [49], however, it is possible that some firms interact with several other firms while others do not have any business partners at all. Furthermore, the extent to which a firm is influenced by its business partners can differ. All these features make the model of [49] much more realistic than the one of [69].

The micro-structural interdependence can be consistently included in the macro-structural model (15.5) as follows

$$r_n = \beta_n(X) \cdot \sqrt{1 - w_{k(n)}^2} \cdot X_{k(n)} + \alpha_n(\varepsilon) \cdot w_{k(n)} \cdot \varepsilon_n(\mathcal{G}, X, \varepsilon) \qquad (15.6)$$

where the weights $\beta_n(X)$ and $\alpha_n(\varepsilon)$ must be determined such that the following assumption holds.

Assumption 15.7.4
(i) $\alpha_n(\varepsilon) \cdot \varepsilon_n(\mathcal{G}, X, \varepsilon)$ *is normalized to standard normal.*
(ii) *The marginals of* $(r_1, \ldots, r_N)$ *remain standard normal.*

---
[12] Compare [49], Definition 3.

15.7 Contagion Through Macro- and Microstructural Channels

As there is no canonical way to specify the micro-structural dependence in the idiosyncratic term $\varepsilon_n(\mathcal{G}, Y, \varepsilon)$, [49] suppose that $(r_1, \ldots, r_N)$ has a direct linear effect on $\varepsilon_n(\mathcal{G}, Y, \varepsilon)$, i.e.

$$\varepsilon_n(\mathcal{G}, Y, \varepsilon) = \sum_{m=1}^{N} \xi_{nm} r_n + \eta_n \varepsilon_n.$$

Then the asset return of firm n is given in latent factor representation by

$$r_n = \beta_n(X) \cdot \sqrt{1 - w_{k(n)}^2} \cdot X_{k(n)} + \alpha_n(\varepsilon) \cdot w_{k(n)} \cdot \left(\sum_{m=1}^{N} \xi_{nm} r_n + \eta_n \varepsilon_n \right).$$

Note that this equation reduces to equation (15.6) when we set all $\xi_{nm} = 0$ and $\eta_n = 1$. In this case, $\alpha_n(\varepsilon) = 1$ in order for Assumption 15.7.4 (i) to hold, as ε_n is already standard normal. Condition (ii) in Assumption 15.7.4 is satisfied when $\beta_n(X) = 1$. [49] discuss various methods to derive mathematically rigorous conditions for Assumption 15.7.4 using recursive integration methods.

15.7.2 The Rating Migrations Process

The model introduced in [49] focuses on the process of rating migrations of firms in a given portfolio. Suppose that the creditworthiness of a debtor n at time t is fully described by its state variable or rating S_t^n taking values in the set of rating categories $\mathcal{R} = \{1, \ldots, R\}$. Here, 1 denotes the default state. In [49] it is assumed that the joint rating dynamics $S_t \in \mathcal{R}^N$ follow a discrete-time, stationary Markov chain $(S_t)_{t \in \mathbb{N}} = (S_t^1, \ldots, S_t^N)_{t \in \mathbb{N}}$. The transition probabilities between the different rating classes are summarized in a transition matrix T, where the transition from a state $s = (s_1, \ldots, s_N)$ at time t to a state $q = (q_1, \ldots, q_N)$ at time $t+1$ is given by

$$T_{sq} = \mathbb{P}(S_{t+1} = q | S_t = s).$$

Note that, due to the stationarity of the Markov chain $(S_t)_{t \in \mathbb{N}}$, the transition rate T_{sq} is independent of the time t. It only depends on the size of the time step which is fixed to one, i.e. we only consider transition probabilities from one time step to the next.

[49] express the macro-structural dependence between the obligors in the portfolio by the common dependence of the rating dynamics on an underlying risk factor X. Conditional on X, ratings are assumed to be independent.

Assumption 15.7.5
The rating dynamics of the different firms in the portfolio are independent conditional on some random vector X.

The conditional independence structure has the convenience that the transition matrices have a product structure, i.e.

$$T_{sq} = \mathbb{E}\left[\prod_{n=1}^{N} T^n_{s_n q_n} \bigg| X\right] \qquad (15.7)$$

where

$$T^n_{s_n q_n}|X = \mathbb{P}\left(S^n_{t+1} = q_n | S^n_t = s_n, X\right)$$

is the individual conditional transition matrix of debtor n. Moreover, as each obligor is only influenced by the sector factor X_k of the sector to which it belongs, equation (15.7) can be further simplified to

$$T_{sq} = \mathbb{E}\left[\prod_{n=1}^{N} T^{k(n)}_{s_n q_n} \bigg| X\right], \qquad (15.8)$$

where $k(n)$ denotes the sector to which obligor n belongs. Using this framework, one can express the conditional migration matrices $T^k|X$ in terms of a latent variable model. Therefore, we associate with each rating class s, some threshold values

$$-\infty = b_{s,1} \leq b_{s,2} \leq \ldots \leq b_{s,R} \leq b_{s,R+1} = \infty.$$

Hence, obligor n migrates from a rating class s at time t to a new class q at time $t+1$, when the synthetic asset value r_n of firm n lies between the threshold values $b_{s,q}$ and $b_{s,q+1}$, i.e. when

$$T^n_{sq} = \mathbb{P}\left(S^n_{t+1} = q | S^n_t = s\right) = \mathbb{P}\left(r_n \in [b_{s,q}, b_{s,q+1}]\right).$$

Note that Assumption 15.7.5 translates to Assumption 15.7.1 when switching from the rating migrations process $(S_t)_{t \in \mathbb{N}}$ to the asset value representation.

15.7.3 Results and Discussion

[49] apply their model to several credit portfolios each consisting of 102 obligors, where counterparties are grouped into $K = 14$ different industry sectors according to the sector classification scheme from the Swiss agency BAK.[13] They use historical default rates for the different sectors which are derived from a time series between 1980 and 1997. The migration matrix is obtained from Moody's based on historical default data between 1970 and 2002. Given the default rates and the migration matrix for 8 different rating classes, [49] calibrate the thresholds (b_{sq}), the risk weights (w_k), and the correlation matrix Σ. Note that, while the macro-structural parameters Σ, w_k and b_{sq} of

---
[13] Konjunkturforschung Basel AG

15.7 Contagion Through Macro- and Microstructural Channels

the latent variable framework can be statistically calibrated from historical default data, the micro-structural parameters Ξ and η need to be determined by expert knowledge.

Applying Monte Carlo simulation the authors analyze the effect of different interdependence structures on the rating distributions for portfolios of different credit qualities and a time horizon of 5 years. The simulations show that higher correlation leads to an increase in the speed of migration through the different categories of the transition matrix which eventually has a significant impact on the volatility, and hence on the credit risk calculations. Similar results have also been derived in the other default contagion models we discussed in these lecture notes.

[49] examine also the effects of the rating-count volatility on the default risk and the loss distribution. They use both VaR and Expected Shortfall as risk measures and focus on the probability distribution of potential credit losses. The results show that micro-structural interdependence leads to an increase in the corresponding risk figures. The authors show that micro-structural interdependencies significantly increase the correlation among debtors and fattens the tails of the credit loss distribution. Since business interdependencies are mostly tail effects, the findings hint at the danger of using Value-at-Risk to manage credit risk and give strong support for tail-sensitive risk measures such as Expected Shortfall.

Compared to previously discussed models for default contagion, the approach of [49] allows for asymmetric relationships between business partners which is more realistic as, for example, a large company might be less affected by the default of a smaller business partner then vice versa.

A potential drawback of this approach, however, is that the calibration process is quite tedious and can be numerically unstable when using the recursive integration method. The suggested approximated recursive integration model converges to the recursive integration model, however, the speed of convergence is not discussed in the paper of [49].

16
Equilibrium Models

The prevalence of financial crises suggests that the financial sector is very sensitive to economic shocks in the sense that shocks which initially only affect a particular region can spread by contagion to the whole financial sector. One approach to model the propagation of contagion has been introduced in [6]. The authors concentrate on a single channel of financial contagion, namely the region-overlapping claims of an interbank deposit market, and model financial contagion as an equilibrium phenomenon. They focus mainly on the overlapping claims that different regions or sectors of the banking system have on one another. These interregional claims provide an insurance against liquidity preference shocks. The drawback, however, is that a small liquidity preference shock or even a bank crisis can spread from one region to the other regions which then might suffer a loss because their claims on the troubled region fall in value. If this spill-over effect is strong enough, the initially small shock in one region can cause a crisis in the adjacent regions and in extreme cases, the crisis can even pass from region to region and become contagious. [6] provide a micro-economically founded model of a banking system and consumers with uncertain liquidity preferences. Their analysis includes a first-best allocation and its decentralization. In their central result they provide sufficient conditions for an economic-wide crisis in a perturbed model. Furthermore they analyze, by illustrating different patterns of cross holdings, how completeness and connectedness of the interbank market structure can channel or attenuate contagion. A general problem with this approach, however, is that the additional uncertainty arising from the interaction of firms can easily be eliminated by means of diversification. This contradicts in some sense the idea of contagion since the risk, resulting from interactions between firms, is *intrinsic* and, thus, actually cannot be diversified away.

In this chapter we review the recent work of [80] which provides an interactive equilibrium model of credit ratings and which does not face the above mentioned shortcoming. [80] assume that the ratings of individual obligors in a portfolio can be affected by small external shocks which might be particular to only a small number of obligors in the portfolio. Due to counterparty relationships in form of borrowing or lending contracts, however, the downgrade of one firm can lead to further downgrades of other firms. Thus an external shock can provoke a cascade of further downgrades. This mechanism can be understood as an infectious chain reaction that spreads from a small group of firms to the whole economy. [80] provide a method to quantify the additional risk that arises from dependencies between borrowers and which can even be computed analytically. In contrast to several other approaches already discussed in these lecture notes, the approach of [80] has the advantage that the risk arising from dependencies between different obligors cannot be diversified away. For practical applications this feature is very realistic. Finally, the approach allows to identify the source of large portfolio losses a priori.

The remainder of this chapter is structured as follows. First, we present in Section 16.1 the setup of a stochastic mean-field as introduced in [80]. The downgrade cascade is defined in this setting and first results on the asymptotic behavior of the total number of downgrades are provided. Afterwards, in Section 16.2 the model is extended to the case of both local and global interactions. Furthermore, we study in Section 16.3 the influence of the downgrade cascade on large portfolio losses and determine the asymptotic loss distribution of a portfolio.

16.1 A Mean-Field Model of Credit Ratings

We start with a description of the simplest case of a downgrade cascade as introduced in [80]. That is, we neglect in this section any possible local interactions between different companies and focus only on the influence of global shocks on the individual credit ratings and the feedback effects from deteriorations in the average rating. Therefore, let us consider a portfolio consisting of N heterogeneous firms indexed by $n = 1, \ldots, N$. We denote the state of company n at time $t \in \mathbb{N}$ represented by its credit rating by S_t^n. The downgrade mechanism of credit ratings acts in a discrete manner, such that credit ratings S_t^n take values in a finite subset Λ of $\{0, \pm\lambda, \pm 2\lambda, \ldots\}$. The parameter λ indicates the coarseness of the credit rating adjustments. The smaller λ the finer is the rating classification scheme. A rating is considered the worse, the higher its value is. Thus if $m\lambda$ is the highest rating in the finite set Λ for some $m \in \mathbb{N}$ then this rating class corresponds to the default state. All the other rating categories in Λ represent non-default states.

Denote by ε_n the random variable representing the idiosyncratic parameters that affect the financial situation of firm n. Assume that the idiosyncratic variables $\varepsilon_1, \ldots, \varepsilon_N$ of different firms are independent. For simplicity we start with a *mean-field model* of credit ratings. This means that besides obligor specific characteristic, the credit rating of every individual firm is affected by the

16.1 A Mean-Field Model of Credit Ratings

average rating in the portfolio. Moreover, there are assumed to be no direct business links. The average rating at time t is defined as

$$\bar{S}_t := \frac{1}{N} \sum_{n=1}^{N} S_t^n.$$

In the simplest setup, the individual ratings S_t^n depend linearly on both the average rating $\bar{S}_t$ as well as on the firm specific quantities ε_n. This yields the following structure of credit rating configurations

$$S_t^n = \alpha \bar{S}_t + \varepsilon_n - (\alpha \bar{S}_t + \varepsilon_n) \bmod \lambda. \tag{16.1}$$

The weight $\alpha \in [0,1]$ specifies the strength of the credit interactions represented by the dependence on the average rating (as there are no other interactions in this mean-field framework). The notation "$y \bmod \lambda$" is the usual *modulo* notation; meaning the remainder of the division of y by λ. This structure ensures that individual credit ratings are positive multiples of λ such that the ratings take values in the set Λ of possible ratings. In case $\alpha = 0$, credit ratings of different firms are independent as they are only influenced by the obligor specific quantities ε_n.

The model specification (16.1) allows that credit deteriorations of some companies can spread among the whole economy by means of changes in the average credit rating. For simplicity we assume that the idiosyncratic parameters ε_n are uniformly bounded in which case we can choose a finite set of rating categories. Formally, [80] impose the following assumption.[1]

Assumption 16.1.1
The random variables $(\varepsilon_n)_{1 \leq n \leq N}$ are independent and uniformly distributed on the interval $[0, m\lambda]$ for some $m \in \mathbb{N}$.

Under this assumption it is possible to prove the existence of the equilibrium configurations of credit ratings given by equation (16.1). The following result is due to [80], Lemma 2.2, to which we also refer for the proof.

Lemma 16.1.2 *Fix some time $t \in \mathbb{N}$. Suppose that the interaction between different firms is weak enough in the sense that $\alpha < 1$. If Assumption 16.1.1 is satisfied, then there exists a finite set*

$$\Lambda \subset \{0, \pm\lambda, \pm 2\lambda, \ldots\}$$

and a configuration of credit ratings $(S_t^n)_{1 \leq n \leq N} \in \Lambda$ that satisfies (16.1).

The idiosyncratic factors ε_n are assumed to be unobservable while their distributions are supposed to be observable. Furthermore, assume that credit

---
[1] Compare Assumption 2.1 in [80].

ratings are the only directly observable quantities. In this sense firm specific factors remain private information. Thus, risk managers do not have complete information about the ability of a company to absorb additional financial distress spread in the economy. [80] define the ability of firm n to absorb external financial distress at time $t \in \mathbb{N}$ as

$$b_t^n := (\alpha \bar{S}_t + \varepsilon_n) \bmod \lambda, \tag{16.2}$$

and call this its *threshold level* or *buffer variable*. The smaller the buffer variable, the larger is the distance to the next higher and, thus, worse rating class. Since the distribution of ε_n is known, it is possible to determine the conditional distribution of b_t^n given an equilibrium configuration $(S_t^n)_{1 \leq n \leq N}$. Consider, for example, the "binary choice" case where $\varepsilon_n \in (0, 1)$ and $\lambda = 1$, i.e. there exist only two rating classes 0 and 1 and, thus, a financial institution only knows whether a firm is bankrupt or not. In this case, for fixed time $t \in \mathbb{N}$, the buffer variables are conditionally independent and identically distributed. In the case where $\lambda < 1$ this does not necessarily hold.

We now turn to the process of downgrade cascades initiated by external shocks. Consider a random variable Z which shall represent the size of an external economy wide shock triggering the downgrade cascade. Since individual firms are usually affected to a different extent by the same shock, we define a sequence $(X_n)_{1 \leq n \leq N}$ of random variables satisfying the following condition.[2]

Assumption 16.1.3
The random variables X_n are non-negative, bounded, and, conditional on the macro variable Z, independent and identically distributed with conditional probability distribution function $F(Z; \cdot)$ and conditional mean $\mathbb{E}[X_n | Z] = Z$.

Then we can define the impact of an external economy wide shock of level Z on firm n as

$$Y_n := \frac{X_n}{N}.$$

For large portfolios we obtain from the Law of Large Numbers that

$$\mathbb{P}\left(\lim_{N \to \infty} \sum_{n=1}^{N} Y_n = \mathbb{E}[X_n | Z] = Z \Big| Z\right) = 1.$$

Remark 16.1.4 If we compare this framework with the framework of a Bernoulli mixture model, then one can identify the random variable Z with the factor vector in the mixture models setting.

After a shock has occurred, firm n is downgraded only if the impact of the external financial distress cannot be absorbed by its buffer variable b_t^n. Formally, company n is downgraded if

---
[2] Compare Assumption 2.5 in [80].

16.1 A Mean-Field Model of Credit Ratings

$$b_t^n + Y_n > \lambda.$$

As external shocks should initially only affect a small number of firms, [80] assume that in a large economy the number of initial downgrades is approximately Poisson distributed. Individual downgrades, however, deteriorate the economy wide financial climate through the decline of the average credit rating. This, in turn, produces a feedback effect reinforcing the deterioration effect.

Now we can investigate how the downgrade cascade evolves in time. Therefore, we start with an equilibrium configuration (S_0^n, b_0^n) of credit ratings that satisfy equation (16.1), i.e. for $n = 1, \ldots, N$ we have

$$\begin{aligned} S_0^n &= \alpha \bar{S}_0 + \varepsilon_n - (\alpha \bar{S}_0 + \varepsilon_n) \bmod \lambda, \\ \text{and} \quad b_0^n &= (\alpha \bar{S}_0 + \varepsilon_n) \bmod \lambda. \end{aligned} \quad (16.3)$$

At time $t = 0$ a shock Y_n hits company n. Firm n is downgraded if it cannot absorb the shock. In any case, however, the buffer variable of firm n has to be adjusted to the new situation even if the firm is not downgraded. Hence for $t = 1$ we obtain

$$S_1^n = \begin{cases} S_0^n + \lambda & \text{if } b_0^n + Y_n > \lambda \\ S_0^n & \text{otherwise,} \end{cases} \quad (16.4)$$

and

$$b_1^n = b_0^n + Y_n - S_1^n + S_0^n.$$

The average credit rating changes whenever one or more firms are downgraded. This feedback effect might lead to further downgrades or, at least, to changes in the buffer variables of other companies as well. Hence for $t \geq 2$ the rating configurations are given by

$$S_t^n = \begin{cases} S_{t-1}^n + \lambda & \text{if } b_{t-1}^n + \alpha \cdot \Delta \bar{S}_{t-1} > \lambda \\ S_{t-1}^n & \text{otherwise} \end{cases} \quad (16.5)$$

and

$$b_t^n = b_{t-1}^n + \alpha \cdot \Delta \bar{S}_{t-1} - S_t^n + S_{t-1}^n,$$

where $\Delta \bar{S}_{t-1} = \bar{S}_{t-1} - \bar{S}_{t-2}$ denotes the change in the average credit rating between times $t-2$ and $t-1$. As before companies, that are not able to absorb the additional financial distress, will be downgraded.

As we are interested in the effects of default contagion in large portfolios, we are especially interested in the distribution of the total number of downgrades in a portfolio. Therefore, define the stopping time

$$\tau := \inf \left\{ t \in \mathbb{N} : \bar{S}_t = \bar{S}_{t+1} \right\}.$$

Note that $S_t^n = S_{t+1}^n = \ldots$ for each $t \geq \tau$ and for all $1 \leq n \leq N$. Hence, at time τ the impact of the external shock and the feedback effect are minimal, such that the financial situation in the portfolio stabilizes. At this point the downgrade cascade extinguishes such that no further companies are downgraded. The total number of downgrades D_τ and the number of downgrades in period t are given by

$$D_\tau = \frac{1}{\lambda} \sum_{n=1}^{N} (S_\tau^n - S_0^n) \quad \text{and} \quad D_t = \frac{1}{\lambda} \sum_{n=1}^{N} (S_t^n - S_{t-1}^n), \qquad (16.6)$$

respectively. Note that the stopping time τ as well as both D_τ and D_t depend on the number N of companies in the economy. Therefore, we will sometimes write $D_\tau(N)$ to indicate this dependence. [80] prove that the downgrade process $(D_t(N))_{t \in \mathbb{N}}$ can be described asymptotically, for $N \to \infty$, as a branching process and that the number of firms, initially hit by the global shock, is Poisson distributed. Moreover, they prove that the total number of additional downgrades $D_\tau(N)$, resulting from a decline in the average credit rating of the portfolio, can be described asymptotically, for $N \to \infty$, by a random variable D_τ^∞ which follows a compound Poisson distribution depending on the parameters α and λ as well as on the shock variable Z. The stronger the interaction between individual firms, the fatter is the tail of the distribution of D_τ^∞. In case the obligor specific variables ε_n are uniformly distributed on $(0, \lambda)$, [80] prove that the distribution of D_τ^∞ only depends on the strength α of the interaction between different firms. [3]

16.2 The Mean-Field Model with Local Interactions

In this section, local dependencies are incorporate in the previous model reflecting, for example, borrower-lender relations. For simplicity we consider the model in a default-only mode, where firms are either bankrupt or not bankrupt. Formally, we assume that ε_n is uniformly distributed on $(0, 1)$ for all $n = 1, \ldots, N$, and that the rating grid size $\lambda = 1$. Further, assume that the rating of each company n depends on the rating of its neighbor $n + 1$. Let α_g and α_l denote some positive parameters characterizing the strength of the global and local interaction component, respectively. The symmetric configuration $S_t^n \equiv 0$ satisfies the equilibrium condition

$$S_t^n = \alpha_g \bar{S}_t + \alpha_l S_t^{n+1} + \varepsilon_n - (\alpha_g \bar{S}_t + \alpha_l S_t^{n+1} + \varepsilon_n) \bmod \lambda. \qquad (16.7)$$

Hence, in this framework, rating fluctuations of the neighbor $n+1$ of a firm n have a much stronger impact on firm n's own rating S_t^n than changes in the ratings of non business partners $i \in \{1, \ldots, N\} \setminus \{n, n+1\}$. This is due to the fact that downgrades of non business partners only affect firm n's rating

---
[3] For details see [80], Theorem 2.7.

16.2 The Mean-Field Model with Local Interactions

through changes in the average rating $\bar{S}_t$. Note that, for $\alpha_l = 0$, we are back in the mean-field interaction case of the previous section. The following definition of *global* and *local defaults* is due to [80], Definition 2.9.

Definition 16.2.1 (Local and Global Defaults)
Company n defaults *locally* if it gets downgraded because of the insolvency of its business partner, firm $n+1$. The corporation defaults *globally* if the default is due to a deterioration of the overall business climate by the average rating throughout the whole portfolio.

[80] impose the following assumption on the chain reaction process.

Assumption 16.2.2
The chain reaction of defaults is modeled as an alternating sequence of global and local interactions.

Now, we can specify the cascade of downgrades in the framework of both global and local interactions. Initially, let the stochastic processes $\{(S_t^n, b_t^n)\}_{t \in \mathbb{N}}$ of rating configurations and buffer variables be given by

$$S_0^n = 0, \quad \text{and} \quad b_0^n = \varepsilon_n,$$

for $1 \le n \le N$. Similarly to the mean-field case, an external shock hits firm n at time $t = 1$. Then the ratings and buffer variables for each firm n change to

$$S_1^n = \begin{cases} 1 & \text{if } b_0^n + Y_n > 1 \\ 0 & \text{otherwise} \end{cases} \quad \text{and} \quad b_1^n = b_0^n + Y_n - S_1^n.$$

This corresponds to equation (16.4) of the mean-field case since due to the above simplifications we have $\lambda = 1$ and $S_0^n = 0$, that is we started in a non-default situation.

The external shock to firm n can lead to global defaults of some firms due to changes in the average credit rating. These can then trigger local defaults through local interactions defined by equation (16.7). In other words, if firm n defaults, this obviously has a negative effect on firm $n-1$ whose rating S_t^{n-1} depends on S_t^n due to equation (16.7). Hence the downgrading cascade (or in this default-only mode setting the default cascade) can be defined inductively by the following steps.

1. At time $t \in \mathbb{N}$, let the pair (S_t^n, b_t^n) be already defined for $1 \le n \le N$. Denote by
   $$H_t^g := \{1 \le n \le N : S_t^n = 1\}$$
   the set of all firms that have globally defaulted up to period $t \in \mathbb{N}$. Recall that $\lambda = 1$ denotes the default state in this default-only mode setting.

2. We introduce the auxiliary process $\left\{(\tilde{S}_u^n, \tilde{b}_u^n)\right\}_{u \in \mathbb{N}}$ starting in (S_t^n, b_t^n). For $n \in H_t^g$, we set $\tilde{S}_u^n \equiv \tilde{b}_u^n \equiv 1$ for all $u \in \mathbb{N}$. This means that there is no recovery from global defaults. If firm n has not defaulted by time t, it

may default afterwards because of bankruptcy of its partner. Formally, if $n \notin H_t^g$, then we define the auxiliary process at time $u+1$, given its value at time u, by

$$\tilde{S}_{u+1}^n = \begin{cases} 1 & \text{if } \tilde{b}_0^n + \alpha_l \cdot \tilde{S}_u^{n+1} \geq 1 \\ 0 & \text{otherwise} \end{cases}$$

and

$$\tilde{b}_{u+1}^n = \begin{cases} 1 & \text{if } \tilde{S}_u^n = 1 \\ \tilde{b}_0^n + \alpha_l \cdot \tilde{S}_u^{n+1} & \text{otherwise.} \end{cases}$$

Hence, the auxiliary variable $\tilde{S}_u^n$ denotes the rating of firm n at time u after period t (formally, this is a time change where time $u = 0$ corresponds to time t in the old time scale). When the rating $\tilde{S}_u^{n+1}$ of the business partner $n+1$ of firm n equals one for some time $u \in \mathbb{N}$, that is when firm $n+1$ defaults at time u, then also the rating $\tilde{S}_u^n$ of firm n changes to one (meaning that firm n defaults) if it cannot absorb this local shock, that is if $\tilde{b}_0^n + \alpha_l \cdot \tilde{S}_u^{n+1} \geq 1$. Similarly, we have to adjust the auxiliary buffer variable $\tilde{b}_u^n$ of firm n according to a possible default of the business partner $n+1$. The above equation corresponds to equation (16.5) of the mean-field situation when we replace the impact of the change in average rating by the impact of the change in rating of firm n's business partner, and keeping in mind that we are in the default-only mode situation.

3. The set of companies that default because of bankruptcy of its business partner in period t is given by

$$H_t^l := \left\{ 1 \leq n \leq N : n \notin H_t^g \text{ and } \tilde{S}_\infty^n = 1 \right\},$$

where $\tilde{S}_\infty^n = \lim_{u \to \infty} \tilde{S}_u^n$. These local defaults now have a feedback effect on the overall economic conditions for the surviving firms as the average rating changes.

4. To capture this feedback effect, we switch back to the original process $(S_t^n, b_t^n)_{t \in \mathbb{N}}$ and define at time $t+1$ the rating and buffer variable of firm n by

$$S_{t+1}^n = b_{t+1}^n = 1 \quad \text{if } n \in H_t^g \cup H_t^l$$

and by

$$S_{t+1}^n = \begin{cases} 1 & \text{if } \tilde{b}_\infty^n + \alpha_g \cdot \frac{|H_t^l|}{N} \geq 1 \\ 0 & \text{otherwise} \end{cases}$$

and

$$b_{t+1}^n = \begin{cases} \tilde{b}_\infty^n + \alpha_g \cdot \frac{|H_t^l|}{N} & \text{if } S_{t+1}^n = 0 \\ 1 & \text{otherwise,} \end{cases}$$

for the firms that have not defaulted by time t.[4] Here $\tilde{b}^n_\infty := \lim_{u \to \infty} \tilde{b}^n_u$. Note that, in the default-only mode setting, $|H^l_t|/N$ equals the change in the average rating at time t due to local defaults. Here we used that, due to Assumption 16.2.2, there are no global defaults at time $t+1$ as local and global defaults occur alternatingly and we started with a global shock in time t.

Analogously to the mean-field case, we can consider the stopping time $\tau := \inf\{t : \bar{S}_t = \bar{S}_{t+1}\}$, representing the time of extinction of the downgrade cascade. The number of bankruptcies D_t in period t and the total number of defaults D_τ are defined by equation (16.6). In a model of global and local interactions the total number of defaults, triggered by an external shock, can be written as

$$D_\tau = \sum_{t=0}^{\tau} D_t = \sum_{n=1}^{N} S^n_\tau$$

as $S^n_0 = 0$ for all $1 \leq n \leq N$. Similar to the results obtained in the previous section, it is possible to determine the asymptotic distribution of the number of downgrades in the case of local and global interactions. If the interaction is weak in the sense that

$$\frac{\alpha_g}{1 - \alpha_l} \leq 1,$$

then [80] prove that, in the binary choice situation of this section, the sequence of downgrades $\{D_t\}_{t \in \mathbb{N}}$ can be approximated in law by a branching process. Moreover, the total number of defaults $D_\tau(N)$ in a portfolio of N companies converges in distribution to an almost surely finite random variable D^∞_τ as $N \to \infty$.[5] Furthermore, [80] obtain that purely local interactions cannot generate a heavy tailed distribution of defaults if the interactions of the firms are not strong enough.

16.3 Large Portfolio Losses

To demonstrate the impact of local and global interactions and, thus, the impact of the downgrade cascade on large portfolio losses, [80] study portfolios with financial positions whose market values depend on the credit rating of the issuer as, for example, loans or bonds. Changes in the average rating of the portfolio can significantly reduce the market value of such a portfolio. In this section we concentrate on the asymptotic behavior of the portfolio loss as the number of firms increases to infinity. In particular, we will focus on the impact of global and local interactions on the asymptotic portfolio loss distribution. Therefore, we consider the following setting. In case firm n is downgraded, we denote the random loss, the bank experiences on position n,

---

[4] Here $|H^l_t|$ denotes the number of firms in the set H^l_t.
[5] For details see [80], Theorem 2.10.

by L_n. Accordingly, the total loss of a portfolio, consisting of loans to firms $1 \leq n \leq N$, is given by

$$L_N := \sum_{n=1}^{D_\tau} L_n.$$

Assumption 16.3.1
The individual losses are independent and identically distributed according to some exogenously specified distribution function F.[6]

The number of downgrades or defaults is specified by a branching process according to the previously obtained results. Under the above assumption, [80] obtain the following result.[7]

Proposition 16.3.2 *For $N \to \infty$, the random variable L_N converges in distribution to the random variable*

$$L_\infty = \sum_{n=1}^{D_\tau^\infty} L_n,$$

where D_τ^∞ is the limit, as $N \to \infty$, of the total number of defaults $D_\tau(N)$ in the portfolio.

The random variable L_∞ can be interpreted as the loss suffered by a bank due to negative changes in the ratings of the individual obligors in the portfolio. As we are particularly interested in large portfolio losses, special attention must be drawn to the tail of the distribution of L_∞. The individual loss distributions in a portfolio can usually be controlled quite well by an active risk management such that, in order to reduce credit risk, one can eliminate positions with heavy-tailed distributions from a credit portfolio. In this way, the risk inherent in the occurrence of large individual losses can be reduced. Thus, in the management of large portfolio losses, one is particularly interested in aggregate portfolio losses when it is not possible to reduce credit risk by diversification, i.e. in the case of a relatively large number of average losses. In that case we have a relatively large D_τ and $L_1, \ldots, L_{D_\tau}$ close to their expected values. This case is of particular interest as the risk from interactions between firms is *intrinsic* and, thus, cannot be diversified away by an active risk management. The next definition characterizes the tail structure of aggregate portfolio losses based on the distribution of the random variables L_n.[8]

Definition 16.3.3
Let $(L_n)_{n \in \mathbb{N}}$ be a sequence of non-negative iid random variables defined on a probability space $(\Omega, \mathcal{F}, \mathbb{P})$ with distribution function F.

---

[6] Compare [40] for details on this assumptions.
[7] Compare [80], Proposition 3.1.
[8] See Definition 3.2 in [80].

1. We say that the random variable L_n has an *exponential tail* if the moment generating function $M_{L_n}(s) := \mathbb{E}[\exp(sL_n)]$ exists for small enough s.
2. Let $\bar{F}(x) = 1 - F(x)$, for $x \geq 0$, be the tail of the distribution function F and denote by

$$\bar{F}^{*n}(x) := 1 - F^{*n}(x) := \mathbb{P}[L_1 + \ldots + L_n \geq x]$$

the tail of the n-fold convolution of F. Following [50], we say that F has *sub-exponential tail* if

$$\lim_{x \to \infty} \frac{\bar{F}^{*n}(x)}{\bar{F}(x)} = n \quad \text{for some (all)} \quad n \geq 2.$$

To simplify the analysis we only consider the case where the initial shock affects only one company. Then it can be shown that the random variable D_τ^∞, describing the asymptotic behavior of the total number of downgrades, has a Borel-Tanner distribution with parameter ν given by[9]

$$p_k := \mathbb{P}[D_\tau^\infty = k] = \frac{1}{k!} \cdot (k\nu)^{k-1} \cdot e^{-k\nu} \quad \text{for} \quad k \in \mathbb{N}.$$

The tail distribution is then given by

$$q_k := 1 - \mathbb{P}[D_\tau^\infty < k] = \mathbb{P}[D_\tau^\infty \geq k] = \sum_{l \geq k} p_l \quad \text{for} \quad k \in \mathbb{N}.$$

If individual loss distributions are heavy tailed, then the distribution of aggregate losses is also heavy tailed, resulting in large portfolio losses due to large individual losses. Formally, this is stated in the following theorem.[10]

Theorem 16.3.4 *Suppose that the random variables L_n have sub-exponential tails. In the subcritical case $\nu < 1$ where, on average, each downgrade triggers less than one default, we have*

$$\mathbb{P}[L_\infty > x] = \mathbb{P}\left[\max\{L_1, \ldots, L_{D_\tau^\infty}\} > x\right] \quad as \quad x \to \infty$$

and

$$\mathbb{P}[L_\infty > x] = \mathbb{E}[D_\tau^\infty] \cdot \mathbb{P}[L_1 > x] \quad as \quad x \to \infty.$$

In the subcritical case $\nu < 1$, a downgrade triggers on average less than one additional default. Here the sequence $\{q_k\}_{k \in \mathbb{N}}$ converges to zero at an exponential rate. [80] prove that if the distribution function F of the random variables L_n is continuous and if there exists a constant $c > 0$ such that

$$\int_0^\infty e^{cx} dF(x) < \infty,$$

---

[9] Compare [80], Theorem 2.7.
[10] Compare [80], Theorem 3.4.

then the tail structure of the aggregate loss distribution can be specified in terms of the tails of L_n and the average number of downgrades triggered by a single default.[11] In this sense the tail of the distribution of aggregate losses is the heavier, the slower the decay of the tail distribution of the individual losses.

Now consider the critical case $\nu = 1$ where each downgrade triggers on average another downgrade and the distribution of the total number of downgrades is heavy tailed. One can show that, under some technical assumptions, the distribution of aggregate losses has sub-exponential tails and that large portfolio losses typically result from an unusually large number of individual defaults. Hence, even if the individual loss distributions are light tailed, the aggregate loss distribution is fat tailed if the interactions of the credit ratings are strong. This is formalized in the following Theorem.[12]

Theorem 16.3.5 *Suppose that the distribution function F of the random variables L_n satisfies*

$$\liminf_{x \to \infty} \frac{\bar{F}^{*(n+1)}(x)}{\bar{F}^{*n}(x)} \geq a \quad \text{for all} \quad n \in \mathbb{N} \quad \text{and some} \quad a > 1.$$

If $\nu = 1$, then the random variable L_∞ has a sub-exponential distribution. More precisely,

$$\mathbb{P}[L_\infty \geq x] = \mathbb{P}\left[D_t^\infty \geq \frac{x}{\mathbb{E}[L_n]}\right] \quad \text{as } x \to \infty.$$

16.4 Discussion

[80] present an equilibrium model of credit ratings with both global and local interactions. The downgrade cascade of credit ratings is started by an external shock which initially affects only a small group of firms in the portfolio. The external shock impacts the portfolio in the following mechanism. The shock can directly lead to downgrades of individual companies in the portfolio. These can then produce a feedback effect through two channels. First, the downgrades of individual firms can lead to a deterioration of the average credit rating which can then produce a new wave of downgrades and so forth. Secondly, a downgrade of firm n can lead to a downgrade of its business partner $n+1$, having again a feedback effect on the rating of firm n and the average rating. Henceforth, the initial shock can lead to further financial distress in the portfolio through global and local interactions.

---

[11] Compare [80], Theorem 3.5.
[12] Compare [80], Theorem 3.7.

16.4 Discussion 209

[80] show that in the limit, as the number N of firms in the portfolio tends to infinity, the distribution of the number of downgrades converges to the distribution of a branching process. Moreover, the tail distribution of the number of downgrades has a slow exponential decay whenever the interaction of firms is not too strong. Aggregate losses have heavy tailed distributions if individual loss distributions are heavy tailed or if the interactions among companies are too strong. As additional risk due to downgrade cascades cannot be diversified away, neglecting the importance of contagion effects can lead to an underestimation of credit portfolio risk. Therefore, any sensible risk management should take care of this source of risk and an appropriate model for default contagion should reflect the feature that risk due to interactions between firms is intrinsic.

The model of [80] seems to possess most of the features, a sensible model of credit contagion should have. However, there are some shortcomings and questionable aspects of the model. First of all, the model is based on an equilibrium model in which ratings follow a mechanism of equilibrium selection over time. It is questionable whether credit ratings are in general suitable to be modeled as equilibrium. In reality, ratings characterize the credit worthiness of a company and are based on firm specific characteristics and the firm's financial situation.

Moreover, in the present model only shocks are considered which have a negative effect on the portfolio and which lead to a downgrading of the company. Positive shocks and segmented negative and positive shocks are not considered although a negative shock on the oil price, for example, can have a positive effect on the renewable energy industry. Additionally, with the choice of possible rating classes $\Lambda \subset \{0, \pm\lambda, \pm 2\lambda, \ldots\}$, as presented in the model, it is possible that the average rating takes a negative value. This could lead to positive effects which are, however, not accounted for in the model. A more flexible structure where both up- and downgrades are possible would be desirable.

Furthermore, defaults occurring in the first period should not influence the average rating in the second period since these firms should actually disappear from the portfolio and probably should be replaced by some new firms that have been added to the portfolio. Hence the average rating in the second period should probably be computed as the average rating over all non-defaulted firms. Otherwise it is not very surprising that the aggregate loss distribution is heavy tailed as all defaulted firms influence the average ratings of all later time periods while no new firms are considered.

Finally, the simplifying assumption of mean-field interactions is questionable. A richer interaction structure would be desirable. We have studied some approaches with more realistic networks of interacting companies in Chapter 15. These approaches, however, often have the shortcoming that no analytical solutions can be obtained and, thus, that one has to rely on numerical simulations.

A
Copulas

This short summary of copulas is based on [103] to which we also refer for more details. When modeling credit portfolio risks, we need to take care, on the one hand, of the individual loss distributions L_n of every single obligor n in the portfolio and, on the other hand, of the dependencies between the default events of different obligors. We denote by

$$F_n(x_n) = \mathbb{P}(L_n \leq x_n)$$

the probability distribution function of the loss variable L_n of obligor n. To account for default dependencies in the portfolio, we are interested in the joint distribution function

$$F(x) = \mathbb{P}(L_1 \leq x_1, \ldots, L_N \leq x_N), \quad x = (x_1, \ldots, x_N),$$

with marginal distributions

$$F_n(x_n) = F(\infty, \ldots, \infty, x_n, \infty, \ldots, \infty).$$

In case the individual loss variables are normally distributed, such a joint distribution function can easily be obtained as a multivariate normal distribution function. For arbitrary distribution functions F_n, however, it is not clear a priori how F should be obtained.

Copulas present a method to solve this problem. The joint distribution function of $L_1, \ldots, L_N$ contains information about both the marginal distributions of the individual obligor losses L_n and the dependencies between them. Copula functions present a method to isolate the information on the precise dependence structure of such a random vector. Hence, they are more appropriate to describe complex dependence structures of random variables than marginal distribution functions together with the linear correlation matrix which often fails to capture important risks.[1] Formally, the following definition holds.[2]

---

[1] See [51] for details.
[2] Compare [103], Definition 5.1.

A Copulas

Definition A.0.1 (Copula)
A d-dimensional *copula* $C(u) = C(u_1, \ldots, u_d)$, is a multivariate distribution function on $[0,1]^d$ with standard uniform marginal distributions.

Hence a copula is a mapping $C : [0,1]^d \to [0,1]$ with the following three properties

(1) $C(u_1, \ldots, u_d)$ is increasing in each component x_i.
(2) $C(1, \ldots, 1, u_i, 1, \ldots, 1) = u_i$ for all $i \in \{1, \ldots, d\}$ and $u_i \in [0,1]$.
(3) For all $(a_1, \ldots, a_d), (b_1, \ldots, b_d) \in [0,1]^d$ with $a_i \leq b_i$, it holds

$$\sum_{i_1=1}^{2} \cdots \sum_{i_d=1}^{2} (-1)^{i_1 + \ldots + i_d} \cdot C(u_{1i_1}, \ldots, u_{di_d}) \geq 0$$

where $u_{j1} = a_j$ and $u_{j2} = b_j$ for all $j \in \{1, \ldots, d\}$.

The first two properties are obvious for any multivariate distribution function and its uniform marginal distributions. The third property, the so-called *rectangle inequality,* ensures the non-negativity of the joint probability.

Using copula functions, information on the marginal distributions is neglected. By Sklar's Theorem (Theorem A.0.2 below), however, it can be shown that multivariate distribution functions can be separated into marginal distribution functions and a copula function and, conversely, that copulas and marginal distribution functions can be linked to specify a multivariate distribution function. Hence, the problem of modeling the dependence structure in a credit portfolio can be separated into two steps (identifying marginal distribution functions and choosing an appropriate copula function).

Theorem A.0.2 (Sklar's Theorem) *Let F be a joint distribution function with continuous margins $F_1, \ldots, F_d$. Then there exists a unique copula function $C : [0,1]^d \to [0,1]$ such that for all $x_1, \ldots, x_d$ in $\bar{\mathbb{R}} = [-\infty, \infty]$ the relation*

$$F(x_1, \ldots, x_d) = C(F_1(x_1), \ldots, F_d(x_d)) \tag{A.1}$$

holds. Conversely, if C is a copula and $F_1, \ldots, F_d$ are univariate distribution functions, then the function F given by (A.1) is a joint distribution function with margins $F_1, \ldots, F_d$.

For a proof of this theorem we refer to [116] or [106]. The theorem states that for any multivariate distribution, the univariate margins and the dependence structure can be separated. The latter is completely characterized by the copula function C. Thus the copula is a distribution function $C : [0,1]^d \to [0,1]$ which combines uniform random numbers $U_1, \ldots, U_d$ to a joint distribution function, given by

$$C(u_1, \ldots, u_d) = \mathbb{P}(U_1 \leq u_1, \ldots, U_d \leq u_d).$$

A Copulas 213

In case of continuous marginal distributions, one can introduce the notion of the copula of a distribution as follows.[3]

Definition A.0.3 (Copula of a Distribution)
If the random vector X has joint distribution function F with continuous marginal distributions $F_1, \ldots, F_d$, then the *copula of F* (or X) is the distribution function C of $(F_1(X_1), \ldots, F_d(X_d))$.

In comparison to linear correlation coefficients, copulas have the advantage of being invariant with respect to strictly increasing transformation $T_1, \ldots, T_d$, which means that $(X_1, \ldots, X_d)$ and $(T_1(X_1), \ldots, T_d(X_d))$ have the same copula.[4]

Example A.0.4 (Gauss Copula) Let $Y_1, \ldots, Y_d$ be normally distributed random variables with means $\mu_1, \ldots, \mu_d$, standard deviations $\sigma_1, \ldots, \sigma_d$ and correlation matrix R. Through strictly increasing transformations we obtain new random variables $X_1, \ldots, X_d$ defined by

$$X_i := \frac{Y_i - \mu_i}{\sigma_i}, \quad i = 1, \ldots, d.$$

Hence both random vectors X and Y have the same copula which is given by

$$C_R^{Ga}(u) = \mathbb{P}(\Phi(X_1) \leq u_1, \ldots, \Phi(X_d) \leq u_d) \tag{A.2}$$
$$= \Phi_d(\Phi^{-1}(u_1), \ldots, \Phi^{-1}(u_d); R),$$

where Φ denotes the standard univariate normal distribution function and $\Phi_d(\cdot; R)$ denotes the joint distribution function of X which is a d-dimensional normal distribution with correlation matrix R. In two dimensions and for $\varrho = \mathrm{Corr}(X_1, X_2)$ with $|\varrho| < 1$, the Gaussian copula $C_\varrho^{Ga}(u_1, u_2)$ is given by

$$C_\varrho^{Ga}(u_1, u_2) = \int_{-\infty}^{\Phi^{-1}(u_1)} \int_{-\infty}^{\Phi^{-1}(u_2)} \frac{\exp\left\{\frac{-(s_1^2 - 2\varrho s_1 s_2 + s_2^2)}{2(1-\varrho^2)}\right\}}{2\pi(1-\varrho^2)^{1/2}} ds_1 ds_2.$$

A variant of the Gaussian copula is the t-copula which we will discuss in the next example

Example A.0.5 (t-Copula) Let $Y_1, \ldots, Y_d$ be standard normally distributed random variables with correlation matrix R. Let S be a χ_ν^2-distributed random variable with ν degrees of freedom which is independent of $Y_1, \ldots, Y_d$. Then the random variables

$$X_i := \frac{\sqrt{\nu}}{\sqrt{S}} \cdot Y_i, \quad i = 1, \ldots, d,$$

---

[3] See [103], Definition 5.4.
[4] For a proof of this assertion see [103], Proposition 5.6.

are t_ν-distributed with correlation matrix R where t_ν is the univariate Student t-distribution function with ν degrees of freedom. The distribution depends on ν, but not on the mean or the standard deviation which makes the t-distribution and also the t-copula important in both theory and practice. Thus the random vector $X = (X_1, \ldots, X_d)$ is $t^d_{\nu,R}$-distributed where $t^d_{\nu,R}$ denotes the d-dimensional multivariate t-distribution with ν degrees of freedom and correlation matrix R. By Definition A.0.3 the copula of X is then given by

$$C^t_{\nu,R}(u_1, \ldots, u_d) = \mathbb{P}\left(t_\nu(X_1) \leq u_1, \ldots, t_\nu(X_d) \leq u_d\right)$$
$$= t^d_{\nu,R}\left(t_\nu^{-1}(u_1), \ldots, t_\nu^{-1}(u_d)\right).$$

It is called the *t-copula* with ν degrees of freedom and correlation matrix R and is explicitly given by

$$C^t_{\nu,R}(u) = \int_{-\infty}^{t_\nu^{-1}(u_1)} \cdots \int_{-\infty}^{t_\nu^{-1}(u_d)} \frac{\Gamma\left(\frac{\nu+d}{2}\right)}{\Gamma\left(\frac{\nu}{2}\right) \cdot \sqrt{(\pi\nu)^d \cdot |R|}} \cdot \left(1 + \frac{xR^{-1}x}{\nu}\right)^{-\frac{\nu+d}{2}} dx,$$

where t_ν^{-1} denotes the quantile function of a standard univariate t_ν-distribution.

When modeling extreme credit risk and, especially, concentration risks, we are particularly interested in the tail of the loss distribution. Scatter plots having many points in the upper right or lower left corner indicate high probabilities of extreme events or *tail dependence*. Formally, the following definition holds for pairs of random variables.[5]

Definition A.0.6 (Tail Dependence)
Let X_1 and X_2 be random variables with distribution functions F_1 and F_2, respectively. The coefficient of *upper tail dependence* of X_1 and X_2 is

$$\lambda_u = \lim_{q \to 1-} \mathbb{P}\left(X_2 > F_2^{\leftarrow}(q) | X_1 > F_1^{\leftarrow}(q)\right),$$

provided a limit $\lambda_u \in [0, 1]$ exists. If $\lambda_u \in (0, 1]$, then X_1 and X_2 are said to show upper tail dependence or extremal dependence in the upper tail. If $\lambda_u = 0$, they are *asymptotically independent* in the upper tail. Analogously, the coefficient of *lower tail dependence* is

$$\lambda_l = \lim_{q \to 0+} \mathbb{P}\left(X_2 \leq F_2^{\leftarrow}(q) | X_1 \leq F_1^{\leftarrow}(q)\right),$$

provided a limit $\lambda_l \in [0, 1]$ exists.

The t-copula and the Gaussian copula have very similar properties except for *tail dependence*. While Gaussian copulas are tail independent, meaning that joint extreme events are very unlikely when modeled by a Gaussian copula, t-copulas exhibit both positive lower and upper tail dependence. Thus,

---
[5] See [103], Definition 5.30.

extreme events are more likely when modeled by a t-copula. It can be shown that the parameter ν, representing the degrees of freedom for a t-distribution, controls the degree of tail dependence.[6] Due to the radial symmetry of the t_ν-distribution, the coefficients of lower and upper tail dependence for the t_ν-copula agree and are given by

$$\lambda = 2t_{\nu+1}\left(-\sqrt{\frac{(\nu+1)(1-\varrho)}{(1+\varrho)}}\right),$$

where ϱ is the linear correlation between the two considered random variables.

---

[6] See [103], Example 5.33.

References

1. Acerbi C, Tasche D (2002) On the Coherence of Expected Shortfall. Journal of Banking and Finance 26: 1487-1503
2. Akhavein JD, Kocagil AE, Neugebauer M (2005) A Comparative Empirical Study of Asset Correlation. Working Paper, Fitch Ratings, New York
3. Albrecher H, Ladoucette S, Schoutens W (2007) A Generic One-Factor Lévy Model for Pricing Synthetic CDOs. In: Fu MC, Jarrow RA, Yen J-YJ, Elliott RJ (eds) Advances in Mathematical Finance. Birkhaeuser
4. Aleksiejuk A, Holyst JA (2001) A Simple Model of Bank Bankruptcies. Physica A 299 (1-2): 198-204
5. Allen F, Gale D (1998) Optimal Financial Crises. Journal of Finance 53 (4): 1245-1284
6. Allen F, Gale D (2000) Financial Contagion. Journal of Political Economy 108: 1-33
7. Artzner P, Delbaen F, Eber JM, Heath D (1997) Thinking Coherently. Risk 10 (11): 68-71
8. Artzner P, Delbaen F, Eber JM, Heath D (1999) Coherent Measures of Risk. Mathematical Finance 9 (3): 203-228
9. Asberg Sommar P, Birn M, Demuynck J, Düllmann K, Foglia A, Gordy MB, Isogai T, Lotz C, Lütkebohmert E, Martin C, Masschelein N, Pearce C, Saurina J, Scheicher M, Schmieder C, Shiina Y, Tsatsaronis K, Walker H (2006) Studies on Credit Risk Concentration: an Overview of the Issues and a Synopsis of the Results from the Research Task Force Project. BCBS Publications 15
10. Avesani RG, Liu K, Mirestean A, Salvati J (2006) Review and Implementation of Credit Risk Models of the Financial Sector Assessment Program. IMF Working Paper WP/06/134 (available at http://www.imf.org/external/pubs/ft/wp/2006/wp06134.pdf)
11. Bank M, Lawrenz J (2003) Why Simple, When it can be Difficult? Some Remarks on the Basel IRB Approach. Kredit und Kapital 36 (4): 534-556
12. Basel Committee on Bank Supervision (2001) The New Basel Capital Accord - Second Consultative Document. Bank for International Settlements, Basel
13. Basel Committee on Bank Supervision (2003) The New Basel Capital Accord - Third Consultative Document. Bank for International Settlements, Basel
14. Basel Committee on Bank Supervision (2004) International Convergence of Capital Measurement and Capital Standards: A Revised Framework. Bank for International Settlements, Basel

15. Basel Committee on Banking Supervision (2005) Amendment to the Capital Accord to Incorporate Market Risks. Bank for International Settlements, Basel
16. Basel Committee on Banking Supervision (2005) An Explanatory Note on the Basel II IRB Risk Weight Functions. Bank for International Settlements, Basel
17. Becker S, Düllmann K, Pisarek V (2004) Measurement of Concentration Risk - A Theoretical Comparison of Selected Concentration Indices. Unpublished Working Paper, Deutsche Bundesbank
18. Berger RL, Casella G (2001) Statistical Inference. 2nd Edition, Duxbury Press
19. Billingsley P (1986) Probability and Measure. 2nd Edition, Jon Wiley and Sons, New York
20. Black F, Scholes M (1973) The Pricing of Options and Corporate Liabilities. Journal of Political Economy 81 (3): 637-654
21. Bluhm C, Overbeck L, Wagner C (2003) An Introduction to Credit Risk Modeling. Chapman and Hall, New York
22. Blume LE, Durlauf SN (2000) The Interactions-Based Approach to Socioeconomic Behavior. Wisconsin Madison - Social Systems, Working paper series 1
23. Bonti G, Kalkbrener M, Lotz C, Stahl G (2005) Credit Risk Concentrations Under Stress. Journal of Credit Risk 2 (3): 115-136
24. Burton S, Chomsisengphet S, Heitfield E (2005) Systematic and Idiosyncratic Risk in Syndicated Loan Portfolios. Working Paper, available at http://www.bundesbank.de/vfz/vfz_konferenzen_2005.php
25. Burton S, Chomsisengphet S, Heitfield E (2006) Systematic and Idiosyncratic Risk in Syndicated Loan Portfolios. Journal of Credit Risk 2 (3): 3-31
26. Carling K, Ronnegard L, Roszbach KF (2006) Is Firm Interdependence Within Industries Important for Portfolio Credit Risk? Sveriges Riksbank Working Paper 168
27. Cespedes G, de Juan Herrero JA, Kreinin A, Rosen D (2005) A Simple Multi-Factor 'Factor Adjustment' for the Treatment of Credit Capital Diversification. Working Paper, available at http://www.bundesbank.de/vfz/vfz_konferenzen_2005.php
28. Cespedes G, de Juan Herrero JA, Kreinin A, Rosen D (2006) A Simple Multi-Factor 'Factor Adjustment' for the Treatment of Credit Capital Diversification. Journal of Credit Risk 2 (3): 57-85
29. Cochran WG (1954) Some Methods of Strengthening χ^2 Tests. Biometrics 10 (4): 417-451
30. Credit Suisse Financial Products (1997) CreditRisk+: A Credit Risk Management Framework. Credit Suisse Financial Products, London
31. Crosbie P, Bohn J (2000) Modeling Default Risk. White Paper Moodys/KMV Corp.
32. Crouhy M, Galai D, Mark R (2000) A Comparative Analysis of Current Credit Risk Models. Journal of Banking and Finance 24: 59-117
33. Dai Pra P, Runggaldier WJ, Sartori E, Tolotti M (2008) Large Portfolio Losses; A dynamic Contagion Model. Preprint, arXiv:0704.1348v2
34. Daniels H (1987) Tail Probability Approximations. International Statistical Review 55 (1): 37-48
35. Das SR, Duffie D, Kapadia N, Saita L (2007) Common Failings: How Corporate Defaults are Correlated. Journal of Finance 62 (1): 93-117
36. Das SR, Freed L, Geng G, Kapadia N (2006) Correlated Default Risk. Journal of Fixed Income 16 (2): 7-32

37. Davies BJ (2002) Integral Transforms and Their Applications. 3rd edition, Springer-Verlag, Berlin, New York
38. Davis MAH, Lo V (2001) Infectious Defaults. Quantitative Finance 1: 382-387
39. de Finetti B (1931) Funzione Caratteristica di un Fenomeno Aleatorio. 6. Memorie. Academia Nazionale del Linceo: 251-299.
40. Dembo A, Deuschel J-D, Duffie D (2004) Large Portfolio Losses. Finance and Stochastics 8 (1): 3-16
41. Deutsche Bundesbank (2006) Konzentrationsrisiken in Kreditportfolios. Monatsbericht Juni 2006, Deutsche Bundesbank
42. Diamond D, Dybvig P (1983) Bank Runs, Deposit Insurance, and Liquidity. Journal of Political Economy 91 (3): 401-419
43. Duffie D (2002) A Short Course on Credit Risk Modeling with Affine Processes. Associazione Amici della Scuola Normale Superiore.
44. Duffie D, Eckner A, Horel G, Saita L (2006) Frailty Correlated Default. Working Paper, Stanford University.
45. Duffie D, Saita L, Wang K (2006) Multi-Period Corporate Default Prediction with Stochastic Covariates. FDIC Center for Financial Research Working Paper No. 2006-05
46. Duffie D, Singleton K (2003) Credit Risk: Pricing, Measurement and Management. Princeton University Press, Princeton and Oxford
47. Dufresne PC, Goldstein R, Helwege J (2003) Is Credit Event Risk Priced? Modeling Contagion via Updating of Beliefs. Working paper, University of California at Berkeley.
48. Düllmann K, Masschelein N (2006) Sector Concentration in Loan Portfolios and Economic Capital. Discussion Paper Series 2 Banking and Financial Studies No. 09/2006, Deutsche Bundesbank
49. Egloff D, Leippold M, Vanini P (2004) A Simple Model of Credit Contagion. Maastricht Meetings Paper No. 1142, EFMA 2004 Basel Meetings Paper
50. Embrechts P, Klüppelberg C, Mikosch T (1997) Modeling Extremal Events. Springer-Verlag, Berlin
51. Embrechts P, McNeil A, Straumann D (2002) Correlation and Dependence in Risk Management: Properties and Pitfalls. In: Demster M and Moffatt HK (eds) Risk Management: Value at Risk and Beyond. Cambridge University Press, Cambridge: 176-223.
52. Emmer S, Tasche D (2004) Calculating Credit Risk Capital Charges with the One-Factor Model. Journal of Risk 7 (2): 85-103
53. Encaoua D, Jacquemin A (1980) Degree of Monopoly, Indices of Concentration and Threat of Entry. International Economic Review 21 (1): 87-105
54. Engelage D (2006) A Gibbsian Approach to Market Demand with Locally Interacting Agents. BiBoS Discussion Paper Amendment, E06-08-226
55. Finger C (2001) The One-Factor CreditMetrics Model in the New Basel Capital Accord. Risk Metrics Journal 2 (1): 9–18
56. Focardi SM, Fabozzi FJ (2004) A Percolation Approach to Modelling Credit Loss Distribution Under Contagion. Journal of Risk 7 (1): 79-94
57. Fontes LRG, Sidoravicius V (2004) Percolation. In: Lawler GF (eds) School and Conference on Probability Theory. Trieste, Italy, May 13-17, 2002, Trieste: ICTP - The Abdus Salam International Centre for Theoretical Physics. ICTP Lecture Notes Series XVII: 101-201
58. Frey R, Backhaus J (2003) Interacting Defaults and Counterparty Risk: a Markovian Approach. Working Paper, University of Leipzig

59. Frey R, McNeil A (2001) Modeling Dependent Defaults. Working Paper, ETHZ
60. Frey R, McNeil A (2002) VaR and Expected Shortfall in Portfolios of Dependent Credit Risks: Conceptual and Practical Insights. Journal of Banking and Finance 26: 1317-1334
61. Frey R, McNeil A (2003) Dependent Defaults in Models of Portfolio Credit Risk. Journal of Risk 6 (1): 59-92
62. Frey R, McNeil A, Nyfeler M (2001) Copulas and Credit Risk Models. Risk Magazine :111-114
63. Frey R, Runggaldier W (2007) Credit Risk and Incomplete Information: a Nonlinear-Filtering Approach. Preprint, Department of Mathematics, Universität Leipzig
64. Frye J (2000) Collateral Damage Detected. Federal Reserve Bank of Chicago, Working Paper, Emerging Issues Series 2000: 1-14
65. Gastwirth JL (1972) The Estimation of the Lorenz Curve and Gini Index. The Review of Economics and Statistics 54 (3): 306-316
66. Georgii H-O (1988) Gibbs Measures and Phase Transitions. De Gruyter Studies in Mathematics 9, de Gruyter, Berlin
67. Giesecke K (2004) Correlated Default with Incomplete Information. Journal of Banking and Finance 28: 1521-1545
68. Giesecke K, Weber S (2003) Cyclical Correlations, Credit Contagion, and Portfolio Losses. Journal of Banking and Finance 28 (12): 3009-3036
69. Giesecke K, Weber S (2006) Credit Contagion and Aggregate Losses. Journal of Economic Dynamics and Control 30 (5): 741-767
70. Gordy M (2000) A Comparative Anatomy of Credit Risk Models. Journal of Banking and Finance 24 (1-2): 119-149
71. Gordy M (2002) Saddlepoint Approximation of CreditRisk+. Journal of Banking and Finance 26 (7): 1335- 1353
72. Gordy M (2003) A Risk-Factor Model Foundation for Ratings-Based Bank Capital Rules. Journal of Financial Intermediation 12 (3): 199-232
73. Gordy M (2004) Granularity Adjustment in Portfolio Credit Risk Measurement. In: Szegö G (eds) Risk Measures for the 21st Century. Jon Wiley and Sons
74. Gordy M, Heitfield E (2002) Estimating Default Correlations from Short Panels of Credit Rating Performance Data. Federal Reserve Board Working Paper
75. Gordy M, Lütkebohmert E (2007) Granularity Adjustment for Basel II. Discussion Paper Series 2 Banking and Financial Studies 2007-02-09, Deutsche Bundesbank
76. Gouriéroux C, Laurent JP, Scaillet O (2000) Sensitivity Analysis of Values at Risk. Journal of Empirical Finance 7 (3-4): 225-245
77. Gupton GM, Finger C, Bhatia M (1997) CreditMetrics. Technical Document, J.P. Morgan
78. Hannah L, Kay JA (1977) Concentration in Modern Industry: Theory, Measurement and the UK Experience. Mac Millan Press, London
79. Hohnisch M (2003) Hildenbrand Distribution Economies as Limiting Empirical Distributions of Random Economies. Bonn Econ Discussion Papers No 28/2003
80. Horst U (2007) Stochastic Cascades, Credit Contagion, and Large Portfolio Losses. Journal of Economic Behaviour and Organization 63 (1): 25-54
81. Hull J, White A (2001) Valuing Credit Default Swaps II: Modeling Default Correlations, Journal of Derivatives 8 (3): 12-22
82. Jarrow RA, Yu F (2001) Counterparty Risk and the Pricing of Defaultable Securities. Journal of Finance 56 (5): 1765-1799

83. Jensen J (1995) Saddle-Point Approximations. Oxford University Press
84. Kealhofer S (1995) Managing Default Risk in Derivative Portfolios. Derivative Credit Risk: Advances in Measurement and Management, Renaissance Risk Publications
85. Kiyotaki N, Moore J (1997) Credit Chains. Working Paper, London School of Economics
86. Kocagil AE, Glormann F, Escott P (2001) Moody's RiskCalc for Private Companies: The German Model. Moody's Investors Service, http://www.moodyskmv.com
87. Kosfeld M (2005) Rumours and Markets. Journal of Mathematical Economics 41 (6): 646-664
88. Koyluoglu U, Hickman A (1998) Reconciling the Differences. Risk 11 (10): 56-62
89. Lando D (2004) Credit Risk Modeling: Theory and Applications. Princeton University Press, Princeton, New Jersey
90. Li D (1999) On Default Correlation: a Copula Function Apporach. Working Paper, RiskMetrics Group, New York
91. Liggett TM (1999) Stochastic Interacting Systems: Contact, Voter and Exclusion Processes. Grundlagen für Mathematische Wissenschaften 324, Springer-Verlag, Berlin
92. Liggett TM (2005) Interacting Particle Systems. Reprint of the 1985 Edition, Springer-Verlag, Berlin
93. Liggett TM, Rolles SWW (2004) An Infinite Stochastic Model of Social Network Formation. Stochastic Processes and their Applications 113 (1): 65-80
94. Litterman R (1996) Hot Spots^{TM} and Hedges. The Journal of Portfolio Management 22: 52-75. Special Issue.
95. Lopez JA (2004) The Empirical Relationship Between Average Asset Correlation, Firm Probability of Default, and Asset Size. Journal of Financial Intermediation 13 (2): 265-283
96. Lorenz MO (1905) Methods of Measuring the Concentration of Wealth. Publications of the American Statistical Association 90 (70): 209-219
97. Lucas DJ (1995) Default Correlation and Credit Analysis. Journal of Fixed Income 4 (4): 76-87
98. Markowitz H (1952) Portfolio Selection. The Journal of Finance 7 (1): 77-91
99. Martin R, Thompson K, Browne C (2001) Taking to the Saddle. Risk Magazine 14 (6): 91-94
100. Martin R, Thompson K, Browne C (2001) How Dependent are Defaults? Risk Magazine 14 (7): 87-90
101. Martin R, Thompson K, Browne C (2001) VaR: Who Contributes and How Much? Risk Magazine 14 (8): 99-102
102. Martin R, Wilde T (2002) Unsystematic Credit Risk. Risk Magazine 15 (11): 123-128
103. McNeil A, Frey R, Embrechts P (2005) Quantitative Risk Management – Concepts, Techniques and Tools. Princeton Series in Finance, Princeton University Press
104. McNeil A, Wendin J (2006) Dependent Credit Migrations. Journal of Credit Risk 2 (3): 87-114
105. Merton R (1974) On the Pricing of Corporate Debt: The Risk Structure of Interest Rates. Journal of Finance 29 (2): 449-470
106. Nelsen RB (1999) An Introduction to Copulas. Springer

107. Nirei M (2006) Threshold Behaviour and Aggregate Fluctuations. Journal of Economic Theory 127 (1): 309-322
108. Prahl J (1999) A Fast Unbinned Test on Event Clustering in Poisson Processes. Working Paper, University of Hamburg, submitted to Astronomy and Astrophysics
109. Pykhtin M (2004) Multi-Factor Adjustment. Risk Magazine 17 (3): 85-90
110. Pykhtin M, Dev A (2002) Analytical Approach to Credit Risk Modelling. Risk Magazine 15 (3): 26-32
111. RiskMetrics Group (1997) The Benchmark for Understanding Credit Risk. CreditMetrics Technical Document (available from http://www.riskmetrics.com/research.html).
112. Rosenthal JS (2003) A First Look at Rigorous Probability Theory. 2nd Edition, World Scientifc
113. Schönbucher PJ (2003) Credit Derivatives Pricing Models: Models, Pricing and Implementation. Jon Wiley and Sons, New York
114. Schönbucher PJ (2004) Information-Driven Default Contagion. Preprint, Department of Mathematics, ETH Zürich
115. Schönbucher PJ, Schubert D (2001) Copula Dependent Default Risk in Intensity Models. Working Paper, Bonn University, 2001.
116. Schweizer B, Sklar A (1983) Probabilistic Metric Spaces. New York: North-Holland/Elsevier.
117. Steinherr A (1998) Derivatives. The Wild Beast of Finance. Jon Wiley and Sons
118. Taistra G, Tiskens C, Glüder D (2001) Basel II – Auswirkungen auf Typische Mittelstandsportfolien. KfW, Abteilung Volkswirtschaft
119. Tasche D (1999) Risk Contributions and Performance Measurement. Working Paper, Technische Universität München (available from http://citeseer.nj.nec.com/tasche99risk.html)
120. Vasicek O (2002) Loan Portfolio Value. Risk Magazine 15 (12): 160-162
121. Wilde T (2001) IRB Approach Explained. Risk Magazine 14 (5): 87-90
122. Wilde T (2001) Probing Granularity. Risk Magazine 14 (8): 103-106
123. Wilson T (1997) Portfolio Credit Risk (I). Risk Magazine 10 (9): 111-117
124. Wilson T (1997) Portfolio Credit Risk (II). Risk Magazine 10 (10): 56-61
125. Yu F (2005) Default Correlation in Reduced-Form Models. Journal of Investment Management 3 (4): 33-42

Index

actuarial model, 53
ad-hoc measures, 67–73
All SNC portfolio, 141
analytical approximation, 28, 108, 132
arrival risk, 4
asset-value models, *see* latent variable models
asymptotic portfolio, 34
Asymptotic Single Risk Factor model, 31–41, 76
asymptotically independent functions, 214

Basel Accords, 10, 11
Basel Committee on Banking Supervision, 9, 10, 31
Basel I, 10
Basel II, 11, 31
Borel-Tanner distribution, 207
buffer variable, 200
business matrix, 191

capital diversification
  factor, 120–129
  index, 123
coherent risk measures, 14
concentration
  curve, 68
  ratio, 68
  risk, 63–66
concentration risk, 65
conditional independence
  assumption, 151, 155, 193
  structure, 50, 191

configuration space, 175
copula
  Archimedean, 167
  asymmetric, 168
  Clayton, 167, 168
  function, 165, 211–215
  Gaussian, 167, 168, 213
  models, 152, 165–171
  of a distribution, 213
  t, 167, 168, 213
counterparty risk, 151, 156
counting process, 156
credit
  migrations, 22
  risk, 3
  worthiness, 3
CreditMetrics model, 30, 90, 167
CreditRisk+ model, 53–60, 81, 104
CRSP/Compustat, 159
cumulant generating function, 94
cyclical default dependence, 151, 156

de Finetti's Theorem, 46
default
  clustering, 152, 155
  contagion, 65, 151–154
  correlation, 5, 151, 165
  dependence, 5, 151
  indicator, 6
  probability, 4, 6
    conditional, 26
    in mixture models, 45
    in the CreditRisk+ model, 59

224   Index

in the IRB model, 35
in the Merton model, 22, 25, 26, 29
joint, 6
marginal, 6
risk, 3
default-rate models, *see* reduced-form models
distance-to-default, 30
doubly stochastic
  assumption, 155, 158
  process, 158
  stopping times, 158
downgrade cascade, 200
downturn LGD, 36

economic capital, 5, 17–18, 132
Edgeworth expansion, 93
effective maturity, 37
empirical proportion of low liquidity firms, 180
entropy index
  name concentration, 141
  sector concentration, 142
equilibrium
  distribution, 179
  models, 154, 197–209
equivalence
  of latent variable models, 166
ergodic
  distribution, 178
  process, 178
exchangeable
  model, 7
  vector, 7
Expected Default Frequency, 4, 29, 30
expected loss, 5, 12
  reserve, 12
Expected Shortfall, 15–17
exponential tail, 207
exposure at default, 4, 5
extremal element, 177

factor models, 23, 35
Feller property, 176
Fisher's dispersion test, 161
frailty, 189

generating function, 167
generator, 167

German credit register, 84
Gini coefficient, 70–72
global defaults, 203
Global Industry Classification Standard, 134
granularity adjustment, 75–90, 99, 100, 117

Herfindahl-Hirschman Index, 72, 100, 123
  name concentration, 141
  sector concentration, 142
homogeneous groups, 187

idiosyncratic risk, 5, 25, 65, 75, 140
incomplete information models, 154
independence of loan quantity, 67
infinitely fine-grained, 28, 32, 34, 38, 75
inter-sector correlation, 121
interacting
  intensity models, 152, 153, 173–195
  particle systems, 173
Internal Ratings Based approach, 11, 31
intra-sector correlation, 121
invariant measure, 177
irrelevance of small exposures, 67

KMV model, 29–30, 167

large exposure rules, 100, 138
latent variable, 24
  models, 19, 20, 24, 50, 165
liquidity state
  high, 175
  low, 175
local defaults, 203
Lorenz curve, 69–70
Lorenz-criterion, 67
loss
  expected, *see* expected loss
  exposure, 54
  portfolio, *see* portfolio loss
  unexpected, *see* unexpected loss
  variable, 6
loss given default, 4, 6

macro-structure, 190, 191
marginal sector diversification factor, 125

Index   225

market discipline, 11, 12, 31
maturity adjustment, 37
mean-field
  interactions, 189
  model, 187, 198
Merton model, 19–30, 35
  multi-factor, 23–28, 50, 109, 140
micro-structure, 190, 192
migration risk, 4
minimal capital
  requirements, 11, 31, 37, 38
  standard, 10
mixing distribution, 44, 45
mixture models, 20, 43–51
  Bernoulli, 43
  Poisson, 44
modulo, 199
moment generating function, 58
Moody's Default Risk Service, 159
multi-factor adjustment, 109–120, 133
multi-sector framework, 121, 133

NACE classification scheme, 133
name concentration, 65, 75
nonexplosive process, 156
notional-amount approach, 9

phase transition, 173, 190
point process, 156
Poisson approximation, 44, 56
portfolio
  invariance, 32
  loss, 6
Prahl test, 161
probability
  generating function, 54
  of default, see default probability

rating
  agencies, 4
  migrations, 193
recovery
  rate, 4
  risk, 4
rectangle inequality, 212
reduced-form models, 19, 20, 53, 152, 155, 174
refined ergodic decomposition, 179
regulatory capital, see minimal capital
Revised Framework, 11, 31

risk
  arrival, see arrival risk
  concentration, see concentration risk
  credit, see credit risk
  default, see default risk
  idiosyncratic, see idiosyncratic risk
  migration, see migration risk
  premium, 12
  recovery, see recovery risk
  systematic, see systematic risk
  weighted assets, 10, 37, 87

saddle-point, 94
  approximation, 76, 93–99, 104
sector concentration, 65, 107, 132
semi-asymptotic
  approach, 76, 90–93, 103
  capital charge, 91
Sklar's Theorem, 212
standardized approach, 9, 10
state
  indicator, 6
  space, 175
stochastic intensity, 157
structural models, see latent variable models
sub-exponential tail, 207
superadditivity, 67
supervisory review process, 11, 31
Syndicated National Credits, 140
systematic risk, 5, 25, 65, 75, 140, 151

tail dependence
  lower, 214
  upper, 214
three-pillar framework, 11
threshold, 22, 24
  models, see latent variable models
transfer principle, 67
transition
  matrix, 193
  probabilities, 22, 193
translation invariant measure, 177

unexpected loss, 5, 12
uniform distribution principle, 67

Value-at-Risk, 13–14, 16
voter model, 173
  dynamics, 176

Printed in Germany
by Amazon Distribution
GmbH, Leipzig